Springer Biographies

More information about this series at
http://www.springer.com/series/13617

Jörg Resag

Feynman and His Physics

The Life and Science of an Extraordinary Man

Jörg Resag
Leverkusen, Nordrhein-Westfalen, Germany

ISSN 2365-0613 ISSN 2365-0621 (electronic)
Springer Biographies
ISBN 978-3-319-96835-3 ISBN 978-3-319-96836-0 (eBook)
https://doi.org/10.1007/978-3-319-96836-0

Library of Congress Control Number: 2018951905

This Springer imprint is published by the registered company Springer Nature Switzerland AG
The registered company address is: Gewerbestrasse 11, 6330 Cham, Switzerland

Preface

If you ask people about the most important physicists of modern times, you will hear some names over and over again. Albert Einstein will certainly be there, as will Isaac Newton and Galileo Galilei. Stephen Hawking will also be known to many, for example, through his bestseller “A Brief History of Time” or the television series “The Big Bang Theory.” But where does Richard Feynman stand, the subject of this book?

At the turn of the millennium, the well-known physics portal “Physics World” of the British Institute of Physics (http://physicsworld.com/) has asked for the ten best physicists of all time. Here is the result[1]:

1. Albert Einstein	Special and General Theory of Relativity
2. Isaac Newton	Laws of Motion and Gravitation
3. James Clerk Maxwell	Equations of Electrodynamics
4. Niels Bohr	Quantum Mechanics, Bohr Model of the Atom
5. Werner Heisenberg	Quantum Mechanics, Uncertainty Principle
6. Galileo Galilei	Law of Inertia, Falling Bodies, Refracting Telescope
7. Richard Feynman	Quantum Electrodynamics, Feynman Diagrams
8. Paul Dirac	Quantum Mechanics, Dirac Equation
9. Erwin Schrödinger	Quantum Mechanics, Schrödinger Equation
10. Ernest Rutherford	Rutherford Model of the Atom, Gold Foil Experiment

[1]See, e.g., CERN Courier of Jan 26, 2000, http://cerncourier.com/cws/article/cern/28153.

Other names such as Enrico Fermi, Max Planck, or Michael Faraday would have deserved a place here as well. Stephen Hawking does not appear in the list—maybe it is just too early to assess his contribution to physics correctly.

Most people will probably not be surprised that Albert Einstein is at the top. At the beginning of the twentieth century, he influenced physics like hardly anyone else, revolutionizing our views on the nature of space and time and revealing their deep connection to gravity. His Special and General Theory of Relativity forms the basis of modern physics.

Isaac Newton and James Clerk Maxwell are also clearly among the leaders. They formulated the basic laws of motion, gravitation, and electromagnetism and thus laid the foundations for the whole of classical physics.

Other names are closely associated with the development of quantum mechanics, which challenged our physical worldview in the 1920s and continues to do so today. Along with the theory of relativity, quantum mechanics is the second cornerstone of modern physics – and we will encounter both of them on many occasions in this book.

The most recent physicist in this list is ranked seventh (see Fig. 1). It is Richard Feynman! This position is certainly a dream result if you have to compete with geniuses like Einstein, Newton, or Maxwell.

Richard Feynman (Fig. 2) was one of the most remarkable and famous physicists in the mid- and late twentieth century. On the one hand, this was due to his outstanding achievements in physics, to which we will return in detail in this book. Feynman belonged to the first generation of young

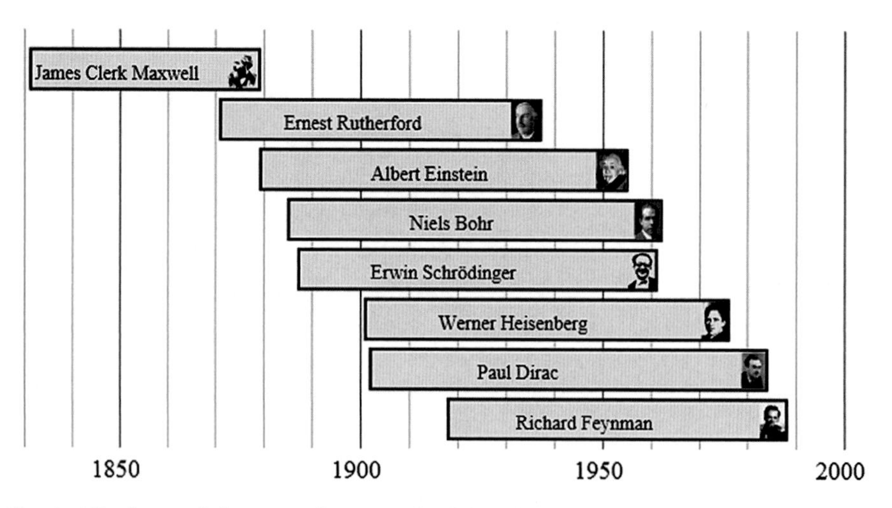

Fig. 1 Lifetimes of the most famous physicists since 1830

Fig. 2 Richard Feynman in 1984 @ Tamiko Thiel, CC-BY-SA 3.0 Unported

physicists who were already familiarized with quantum mechanics during their studies. Prepared in this way, he and some of his colleagues later succeeded in overcoming the enormous difficulties that arose in combining Einstein's Special Theory of Relativity with the principles of quantum mechanics. While his colleagues mostly relied on abstract mathematical formalisms, Feynman followed a more intuitive approach that was typical of him. Based on his own vivid idea of the quantum behavior of particles, he created a completely new formulation of quantum theory which has become the standard tool in relativistic quantum theory today: path integrals and Feynman diagrams. Together with Julian Schwinger and Shinichirō Tomonaga, he received the Nobel Prize for physics in 1965.

But Feynman was not satisfied with that. He was interested in many aspects of physics and was reluctant to commit himself to a single specialty. With his physical intuition, his mathematical abilities, and his deep understanding of quantum theory, he was also able to lead the way in other areas of physics, for example, in the physics of low temperatures (superconductivity and superfluidity). Regarding the so-called weak interaction that triggers, among other things, the radioactive beta decay of atomic nuclei, he and others explained how nature violates the law of reflection symmetry – nature fundamentally distinguishes between right and left! In particle physics, he showed how the data for the scattering of high-energy electrons on protons could be explained by the fact that the electrons were deflected by point-like particles within the protons (partons or quarks).

Throughout his life, Feynman was always interested in computers and the physical foundations of computation, and he later brought quantum mechanics into this field. Today, quantum computers have become an active area of research. He was also one of the first to attempt a quantum description of gravity, i.e., a quantization of Einstein's theory of General Relativity. Even today, this is probably the biggest unresolved problem of physics. A host of physicists are struggling with this, for example, in the context of string theory or loop quantum gravity or as experimenters at large-scale particle accelerators such as the LHC at CERN in Geneva.

But Feynman was not only a great physicist. Much of his fame and popularity can be traced back to his unusual personality. He was an unconventional, cheerful person – more of an eccentric freethinker than the typical, old-fashioned university professor you would often imagine. In addition to his passion for physics, he liked to visit strip clubs and play bongos and later discovered a love for drawing and painting. Unlike some of his colleagues, he had no interest in the emblems of power, whose pompous display he disliked. He also loved getting to the bottom of things and working on a problem until he finally found the solution. He had no problem admitting his own mistakes and hated it when others were not willing to do so, out of vanity or stubbornness. His motto was: "The first principle is that you must not fool yourself – and you are the easiest person to fool."[2]

Feynman was a charismatic speaker and had a talent for dramaturgy, with which he was able to fascinate and inspire his audience. He could captivate you with his humorous and passionate manner, so that in the end you had the feeling of having understood something important – even if you could not always remember exactly what it was.

On the Internet, you can find many videos showing Feynman in action, so that even today you can get an idea of his stirring style of lecture. Bill Gates, the founder of Microsoft, has purchased the BBC videos of seven Feynman lectures and made them freely available to everyone on the Internet at http://research.microsoft.com/apps/tools/tuva/. Feynman gave the lectures in 1964 under the title "The Character of Physical Law" as part of the Messenger Lectures at Cornell University. Take a look at them – the lectures are a real pleasure!

---

[2]See Feynman: *Surely You're Joking, Mr. Feynman!*

Many of Feynman's lectures were also written and published in book form. Even today, his "Feynman Lectures on Physics" from 1961 to 1963 are a treasure trove for every physics student and lecturer. Feynman's deep enthusiasm for physics is clearly visible there. He rethinks all aspects of physics in his own refreshing way, revealing some insights that cannot be found in other standard physics textbooks. No wonder the Feynman lectures are still being printed and purchased, even after more than half a century, something that hardly any other physics textbook can claim.

Feynman's personality predestined him for a task which he tackled with much energy in his later years, at a time when he was fighting against cancer, and which made him known to the general public: his work on the commission to understand the Challenger explosion in January 1986, when all seven crew members of the Space Shuttle died, shortly after the launch. Unlike some of his colleagues on the commission, who spent more of their time in meetings, he went directly to NASA's technicians and engineers and soon discovered the cause of the explosion: a rubber gasket (called an O-ring) that had lost its flexibility in the frosty weather of the launch day, allowing hot gas to escape. The pictures are unforgettable: Feynman dipped such a sealing ring into a glass of ice water in front of the cameras and made it clear to everyone where the problem lay.

Several books have already been written about Richard Feynman's life, along with countless anecdotes – some of which he even wrote himself. This book is therefore not intended to be another comprehensive Feynman biography. Of course, Feynman's life will also play a role in this book and serve as a guide. But the focus will be on what Feynman himself loved so much: physics!

We will try to understand why physics was so fascinating to Feynman and what ideas he and his colleagues actually contributed to it. In doing so, we will focus heavily on one of the fundamental pillars of modern physics, which Feynman quite rightly said nobody understands: quantum mechanics. Of course, this does not mean that we do not have a precise theory of how quantum mechanics works – Feynman himself made important contributions to this. It just means that nobody knows why it works just like that, testing our imagination to the extreme. "But how can it be like that?" asks Feynman in his 1964 Messenger Lectures, illustrating our futile attempt to grasp quantum mechanics with the concepts we are familiar with. Of course, the present book cannot solve this problem, but we can at least try to understand how quantum mechanics works and what is so strange about it.

Our aim is to follow Feynman's path through physics and see how he combined quantum mechanics and relativity theory and how he described antiparticles, the weak interaction, or ice-cold superfluid helium using quantum theory, but also what he thought about nanotechnology and future computers. In doing so, we will encounter a multitude of topics that are fundamental to our current understanding of the laws of nature, and we will thus learn a great deal about the modern worldview provided by physics. We will almost completely dispense with mathematical formulas – sometimes, they are shown in separate boxes and can be skipped without affecting the overall understanding. On the other hand, these boxes will offer the interested reader the opportunity to delve a little deeper into the topic at one point or another.

Feynman would have been one hundred years old in 2018 if his illness had not taken him from us thirty years previously. I hope that, in his honor, I will be able to convey some of the fascination he felt when dealing with physics, which he expressed in the following words from his Messenger Lectures in 1964:

> Our imagination is stretched to the utmost, not, as in fiction, to imagine things which are not really there, but just to comprehend those things which are there.

I would like to take this opportunity to thank Lisa Edelhäuser of Springer Spektrum. It was on her initiative that the German edition of this book about the life and science of Richard Feynman was published again in a new form on his one-hundredth birthday. She worked through the manuscript page by page and made decisive contributions to the success of this complex project with her many constructive suggestions. Bettina Saglio accompanied the finished manuscript to print and discovered many beautiful graphics for the book. Many thanks also to Angela Lahee and Stephen Lyle, who made the English edition of this book possible. And finally, I would like to thank my dear wife Karen and my sons Kevin, Tim, and Jan for their support and patience when the book project took up much more of my time than planned.

Leverkusen, Germany
June 2018

Jörg Resag

Contents

1

Adolescent Years and the Principle of Least Action

These were troubled times when Richard Phillips Feynman was born in New York on May 11, 1918. In Europe, the First World War was still raging and had already cost the lives of countless people. The United States had entered the war about a year earlier (in April 1917), and this would ultimately lead to victory over Germany, although it would still take until November 1918 before the war was brought to an end. Fortunately, Feynman's family was not directly affected by the war.

The early twentieth century was also an eventful time in physics. Albert Einstein (Fig. 1.1) had formulated his Theory of Special Relativity in 1905, fundamentally changing our understanding of space and time. Ten years later (in 1915) he succeeded in demonstrating in his Theory of General Relativity that gravity can be modelled by the curvature of space and time. Even light would be influenced by this curvature and deflected from its straight line by gravity, as he predicted. In May 1919, when a total solar eclipse showed that the light of a star was indeed deflected by the sun's gravitational field, Einstein became famous overnight. Feynman had just turned one year old at the time.

Einstein's Theories of Special and General Relativity were not the only milestones at that time. First results indicated a further revolution in physics, whose full potential would be revealed several years after Feynman's birth and which would become the basis for Feynman's own work: quantum mechanics. A groundbreaking discovery was made by the German physicist Max Planck (Fig. 1.2) in 1900: atoms can only emit or absorb light in certain energy packages (quanta). It was in 1918, the year of Feynman's birth, that Planck was awarded the Nobel Prize for physics for his discovery, which was certainly not a matter of course for a German, given the Kaiser's policy of conquest during the war.

J. Resag, *Feynman and His Physics*, Springer Biographies,
https://doi.org/10.1007/978-3-319-96836-0_1

Fig. 1.1 Albert Einstein (1879–1955) in 1921 (from The Scientific Monthly 12:5, p. 483), https://commons.wikimedia.org/wiki/File:Albert_Einstein_photo_1921.jpg

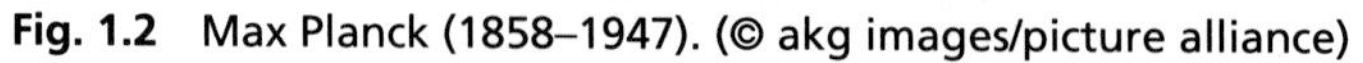

Fig. 1.2 Max Planck (1858–1947). (© akg images/picture alliance)

Planck himself considered the energy quanta to be nothing more than mathematical variables in emission and absorption processes, without any physical reality in themselves. But in 1905 Albert Einstein recognized that these abstract quanta had to be real particles of light, later to be given the name photons. This was the only way to explain, for example, how light in the so-called photoelectric effect could knock out individual electrons from metal surfaces: the photons literally kick them out of the metal.

Light thus consists of particles, but can behave like a wave in many cases. This gives a first taste of the apparent contradictions we have to be prepared for here! Einstein received the Nobel Prize for his photon hypothesis in 1921 – three years after Max Planck. Interestingly, he did not receive the Nobel Prize for his Theories of Special and General Relativity, which made him famous – apparently the Nobel Committee did not consider these revolutionary theories to be sufficiently proven at that time.

So this was the world into which Richard Feynman was born about a century ago and in which he grew up. Feynman had a wonderful childhood and from an early age his father awakened in him a deep enthusiasm for the natural sciences. Not surprisingly, Feynman studied physics at the renowned MIT (Massachusetts Institute of Technology) in Boston. There he encountered two closely related physical principles that are much older than relativity theory and quantum mechanics: Fermat's principle and the principle of least action. They exerted a great fascination on Feynman and would later form the basis for his new view of quantum mechanics.

1.1 Childhood, High School and MIT

Richard Feynman was born on May 11, 1918 in Far Rockaway, a small town in the New York borough of Queens, near the sea on the southern tip of Long Island. Here he spent most of his childhood and youth, which explains his typical New York accent. His physicist colleagues Wolfgang Pauli and Hans Bethe later put it this way: "Feynman spoke like a bum".[1]

Feynman's parents Melville and Lucille were both of Jewish origin. Their families came from Eastern Europe. Melville's parents came from Lithuania and had lived in Minsk (Belarus) before emigrating to the USA in 1895, when Melville was only five years old. Lucille was born that same year in the USA. Her parents had already come to the USA as small children from

[1]See Sykes (1994): *No ordinary genius: the illustrated Richard Feynman.*

Poland and had achieved some prosperity there, so that they were able to acquire a large house in the then still rural Far Rockaway.

For many years this house was also the home of Melville and Lucille with their little son Richard. They shared it with Lucille's sister Pearl and her family for financial reasons. So Feynman grew up with his older cousin Robert and his younger cousin Frances. Lucille gave birth to another son when Richard was five years old, but he died after just one month. It took another four years before Richard's little sister Joan was born on March 31, 1927. Despite the large age difference of nine years, a close bond developed between the two siblings.

Not much is known about Feynman's mother Lucille and her relationship with her son. She must have been a very humorous and warm-hearted woman who, like her son, loved to tell stories. Feynman once said that he learned from his mother that the highest form of knowledge is laughter and compassion. When Feynman was already a famous Nobel Prize laureate and was named *the world's smartest man* by Omni magazine in 1979, she said: "If that's the world's smartest man, God help us". However, she never had much to do with the natural sciences.

Feynman's Father Melville and His Passion for Natural Science

Feynman's father Melville, on the other hand, loved the natural sciences, but had never had the financial means to begin his studies. He wanted to give his son this opportunity, so he did everything he could to inspire him with the secrets of nature and introduce him to scientific thinking. He bought the Encyclopaedia Britannica, put little Richard on his lap, and read to him from it. But he not only read aloud. He also explained what it meant. For example, if it said how big a brontosaur was, he imagined together with Richard what would happen if the dinosaur stood in the front yard of their house. It would be big enough to put its head through the window – only such a huge head would break the window.

Melville also taught his son to keep an independent mind and not be impressed by authorities and their symbols of power. Since Melville was in the uniform business, he knew the difference between the man with the uniform off and the man with the uniform on: it was the same man. In an interview with the BBC in 1981, Feynman tells one of his typical anecdotes in this context[2]:

[2]See also: *The Pleasure of Finding Things Out.*

> One day his father showed him a picture in the New York Times. It showed people bowing in front of the Pope – and Melville was not particularly keen on the Pope. So he asked his son what was so special about this man that made all the others bow before him. And he explained that the difference lay in the hat he wore. Otherwise, he had all the same worries as anyone else: he had to eat and drink and go to the bathroom. He was only human. So it was only his position and his special clothes that made him stand out, nothing particular about what he did or his honourable character.

There are other anecdotes about the way Melville introduced his son to the world and its secrets. Physics was part of this, at least as far as Melville had knowledge of it. One day, for example, the phenomenon of inertia came up. Richard had noticed that a ball would roll to the back of a wagon when the latter was suddenly pulled forward. But Melville showed him that in fact the ball hardly moved – it was actually the back of the wagon that had approached the ball. The young Richard was fascinated, and indeed, if he looked closely, he could see that his father was right. "Why is that?" he asked him. "That nobody knows!" his father replied. "The general principle is that things that are moving try to keep on moving and things that are standing still tend to stand still unless you push on them hard. This tendency is called inertia but nobody knows why it's true!" This was exactly the way in which Feynman himself later approached the mysteries of nature. The term *inertia* alone says little – it's just a name.

Feynman enjoyed these conversations with his father. Many years later, in *The Pleasure of Finding Things Out*, he said: "So that's the way I was educated by my father, with those kinds of examples and discussions, no pressure, just lovely interesting discussions." What a wonderful way to grow up, especially for such an intelligent and inquisitive boy as Feynman.

Joan – Feynman's Talented Sister

Feynman's nine-year-younger sister Joan also developed into a very bright and intelligent child with similar interests to her elder brother. In an intelligence test at high school, she scored 124 points, while her brother scored 123 points – one point less, as Joan jokingly remarked[3]: "So I was actually smarter than he was!" These IQ scores were certainly good, but not exceptional. In later years, Feynman used this result to reject an offer of

[3]See, e.g., Sykes, Christopher: *No Ordinary Genius: The Illustrated Richard Feynman.*

membership from the high IQ society MENSA: he said his IQ was just not high enough. In truth, to Feynman's eyes, this was an elitist club, membership being based on having a certain minimum IQ, and it represented exactly the kind of pompous snobbery he loathed.

It was not easy for Joan to develop her talents. Unfortunately, in the early twentieth century, many still held that the female brain was not naturally suitable for anything as demanding as the natural sciences. Joan's mother Lucille also shared this view. Although from today's point of view it appears completely absurd, it was quite common at that time. And Lucille herself was by no means behind the times in this regard – as a young woman, she had marched for women's suffrage, as Joan's son Charles Hirshberg tells in *My Mother, the Scientist*. When eight-year-old Joan announced that she intended to be a scientist, Lucille replied: "Women can't do science because their brains can't understand enough of it." That came as a major blow to Joan. Her dream to become a scientist seemed to be impossible, and even many years later she still had doubts about her abilities.

Other women also suffered from such prejudices. Some were nevertheless successful. For example, the Austrian nuclear physicist Lise Meitner was involved in the discovery of nuclear fission around 1938, and Marie Curie, a native of Poland, had even received two Nobel Prizes (in physics for the discovery of radioactivity in 1903 and in chemistry for the discovery of the radioactive elements radium and polonium in 1911). In 1918, the year Richard was born, the then 36-year-old German mathematician Emmy Noether found a deep connection between symmetries and conservation laws (Noether's theorem, see Infobox 1.1). Among other things, this answered the question of the origin of inertia, at least in a certain sense. But to Joan, these women seemed to be of another world – and impossible to emulate.

Infobox 1.1: Noether's Theorem and the Origin of Inertia

In 1918, the German mathematician Emmy Noether proved the following fundamental connection:

> Every continuous symmetry of a physical system corresponds to a conservation law.

The term "symmetry" means that we can do something to a physical system without changing its physics. For example, we could move our Solar System to another place in the universe and everything would remain the same, because it doesn't matter where exactly the Solar System is located. The laws of physics are, as far as we know, the same everywhere in the universe. According to Noether's theorem, there must be a corresponding physical quantity that does not change over time and is therefore referred to as a conserved quantity. For the "translation symmetry in space", this conserved quantity is simply the total momentum of the Solar System, i.e., the total momentum of the Sun, planets, and all the other bodies in the Solar System taken together.

Emmy Noether thus taught us that the conservation of momentum is a consequence of the translation symmetry of our world. And conservation of momentum means that there is no change of momentum and therefore no change of velocity without there being some external cause. So our Solar System as a whole would just glide through space at a steady speed and in a fixed direction if there were no external forces acting on it. This is exactly the principle of inertia that Feynman's father explained to him using the example of the ball in the wagon. And Emmy Noether thus discovered a deeper reason for this principle through her theorem.

There are other symmetries and conserved quantities. For example, one could stop the Solar System in thought and let it continue to run later – physics would be the same, because it doesn't depend on the exact moment in time. According to Noether's theorem, it follows that the total energy of the Solar System does not change. Furthermore, the Solar System would not change fundamentally if it were rotated or tilted a few degrees – and so it follows that the so-called angular momentum of the Solar System as a whole remains constant.

For the proof of Noether's theorem one needs the principle of least action, which we will soon encounter. In his popular lecture *Symmetry in Physical Law* in the series *The Character of Physical Law*, Feynman outlines the idea of proof.

So Joan didn't get the same attention from her father as her brother Richard, at least when it came to science. But Richard filled this gap by explaining to his little sister all the wonderful things his father had taught him. Joan would later describe herself as Richard's first student.

One night Richard woke his sister to show her a magnificent light in the sky: it was the aurora borealis! This was a key experience for Joan. It was at this moment that the desire to become an astronomer and deal with celestial phenomena began to take shape in her mind. So Richard gave her an astronomy textbook for her fourteenth birthday, and she worked through it page by page. On page 407, she finally came across a figure that particularly caught her attention. It carried the caption: "Relative strengths of the Mg^+ absorption line at 4481 angstroms … from *Stellar Atmospheres* by Cecilia Payne". There it was: Cecilia Payne! This proved that it was possible: even a woman could become an astronomer!

But it was not easy for Joan to establish herself against all the prejudices then prevalent in science, and she had to fight much harder than her brother to gain a place in this world. But in the end she succeeded, and the solar wind and its interaction with the Earth's magnetic field became one of her main themes. And so it was that the aurora, which Richard had shown her as a child, eventually had its effect. She made a lifelong agreement with her brother: she would leave all other topics to him if only he would leave her the Northern Lights. She wanted to keep this topic all to herself without her clever brother interfering, and he stuck to that agreement.

As Richard grew older, his father's stories were often not enough for him. At about eleven years of age, he set up his own little laboratory at home, and often used it to prepare delicious French fries, as he recounts in *Surely You're Joking, Mr. Feynman*. He loved radios and experimented with their electrical circuits. Eventually, he even earned some money repairing these devices. And his little sister Joan was allowed to join in – Richard paid her 2 cents a week for her help. Apparently, Joan was not only his first student, but also his first assistant.

High School Days

At the age of 13, Feynman went to Far Rockaway High School, which he attended from 1931 to 1935. Like many other gifted children, he was often under-challenged and bored. He learned most not from school lessons, but from books or conversations. Some of the teachers who recognized his talent helped him to do so, for example by lending him advanced math books. Feynman loved these books and his mathematical knowledge was soon far superior to that of his peers. He became the star of the school's **competitive** mathematics group and even won the New York University Math Championship in his last year at high school. In doing so, Feynman benefited from an ability that was also essential for his later success: he did not have to work strictly according to a scheme like many of his fellow students in order to solve a problem. On the contrary, he was not keen on such predefined solutions and always tried to understand and deduce everything from scratch. With his mathematical intuition, he was often able to guess in situations where others were calculating according to some predetermined scheme. Feynman was reluctant to follow fixed rules, even in mathematics.

All in all, his interest was rather one-sided: he loved mathematics and the natural sciences, but had little to do with the humanities, English, religion, or even philosophy. From his point of view, these subjects had little substance. In his youth, Feynman did everything he could to keep contact with these disciplines to a minimum.

In his last year at high school, Feynman was lucky enough to have a young physics teacher who had just arrived at the school: Abram Bader.

He had previously worked on his doctoral thesis with the well-known physicist Isidor Isaac Rabi, but because of the global economic crisis he had run out of money. It was bad luck for Bader, but good luck for Feynman!

Bader realized that Feynman was bored with physics classes. So after one physics lesson he took him aside to introduce him to a particularly interesting physical concept, which unfortunately had no place in normal school lessons: the principle of least action. In Feynman's Lectures on Physics, Feynman deals with this topic in Vol. II, Chap. 19, where he recalls: "Then he told me something which I found absolutely fascinating, and have, since then, always found fascinating. Every time the subject comes up, I work on it."

We will take a closer look at the details of this principle later on, but the basic idea is amazingly simple. Here it is. Imagine a stone moving from one place to another in a gravitational field, which would of course take a certain amount of time. We can calculate its motion step by step using Newton's law if we know its initial speed. However, we can also consider imaginary trajectories between the two places, which take just as long, but do not obey Newton's law of motion. These imaginary motions would therefore not be "chosen" by the stone. Nevertheless, we can still ask ourselves what it would mean if it were to be guided along such a path as if by magic.

We can calculate the kinetic energy at every moment on the path and subtract the potential energy, no matter whether it is a real or imaginary motion. We can then sum these energy differences up (more precisely, we integrate them) over the entire time along the whole path. For each of the motions (imaginary as well as real) we get a number, which is called the *action* of the motion. (This is a somewhat confusing term, but it is just a name for the number we calculate for each motion.) And here it comes: the action we get for the imaginary motions is always bigger than the action for the physical motion, which corresponds to Newton's law. Nature always chooses the motion with the smallest action!

Apparently, we can find the right motion without Newton's law, by searching for the one with the smallest action. This is amazing, because at first glance the two descriptions do not seem to have much in common. It even turns out that all the fundamental laws of nature known today can be described by a suitable action, and this suggests that the principle of least action must have a very fundamental character in nature. But how does nature actually find the path with the smallest action? Does the flying object somehow smell the action of the imaginary paths and then choose the one with the smallest action? Well, as we shall see, this idea is not so far from the truth!

Arline, the Love of His Life

In addition to this experience, which would have a decisive influence on his later scientific work, there was another encounter in Feynman's high school years that would strongly influence his life outside of physics: he got to know Arline Greenbaum (often misspelled "Arlene"). She became the great love of his life, but unfortunately their relationship would come to a tragic end, for Arline died of tuberculosis on June 16, 1945, at the age of only 25 years (Fig. 1.3).

Arline was a pretty girl with long hair who lived not far from the Feynman's. She was very popular with the boys in Far Rockaway and many would have liked to go out with her. In the end, however, it was Richard, who managed to win her heart, even though he was a bit shy as a young man. At first glance, they didn't seem to fit together so well: Arline was cultivated, liked playing the piano, danced, drew, and took an interest in literature and art – all the things that Richard tended to avoid. And yet they were

Fig. 1.3 Richard and Arline (© Emilio Segre visual Archives/American Institute of Physics/Science Photo Library)

soul mates, complementing each other in a wonderful way. They both loved life and faced the world with an unconventional mixture of adventure and open-mindedness. Arline's favorite remark was: "What do you care what other people think?" – a phrase that later became the title of Feynman's last autobiographical book. With this remark she provided Richard with encouragement when he was uncertain and came into conflict with established ideas. He would need this support when he later began to go his own way, and Arline's dictum continued to do its work in Feynman's mind long after she had died.

Going to MIT: Feynman Learns Quantum Mechanics – And We Can Learn with Him

In the summer of 1935, Feynman's high school years came to an end. In almost all subjects he had passed with distinction, even in English, which was not his favorite subject. His parents were determined to give him a college education – an opportunity that Feynman's father Melville had never been fortunate enough to have. However, Columbia University in New York rejected Feynman's application despite his excellent grades, because it had already used up its quota of Jewish students. It is hard to believe that there was such a thing back then, but anti-Semitism was still widespread at that time. Feynman had more luck at MIT (Massachusetts Institute of Technology) in Cambridge, near Boston, and even received a small scholarship of $100 a year.

So in the fall of 1935, at the age of 17, Feynman drove to Boston, about 350 km northeast of New York. Or rather, he was picked up by some fellow students who hoped he would join their student association. A fellow student as talented as Feynman was in great demand, and Feynman was flattered: "It was a big deal; you are grown up!"

Initially, Feynman had enrolled in mathematics at MIT, but he soon realized that this was too theoretical for him. So he changed his mind and tried electrical engineering, but that was too practical. Eventually, he found the golden mean: physics. Here he felt at last that he was in good hands.

During the 17 years since his birth, physics had developed enormously. Within the framework of quantum mechanics, it was finally understood how electrons move in the shells of atoms. Quantum mechanics had thus become the fundamental theory of the subatomic world, and Feynman and his fellow students had the opportunity to get to know this new theory in some detail during their studies. So let's begin by taking a closer look at what quantum mechanics is all about.

The first important finding was that the electrons in the atom do not move on fixed orbits around the atomic nucleus, as proposed by the Danish physicist Niels Bohr in 1913. Rather, they had to be described by waves, as the French physicist Louis de Broglie pointed out in his famous doctoral thesis in 1924. Just as light waves are composed of special particles (photons), electrons are also related to certain electron waves, according to the same formulas:

$$E = h \cdot f$$
$$p = h/\lambda.$$

Particles with high energy E thus belong to waves with high frequency f, where a large particle momentum p leads to a short wavelength λ. The conversion factor between particle and wave properties is given by the Planck constant h – a fundamental physical constant whose value must be determined by experiment. And these relations actually apply quite generally to any object in the quantum world, no matter whether it is a photon, an electron, or a proton.

The most common symbol for the electron wave (which is also often called the *wave function*) is the Greek letter ψ. In many cases it is sufficient to imagine this electron wave as being similar to a water wave. Positive values of ψ stand for a wave peak and negative values for a wave trough.

If one wants to be mathematically correct, the values of the electron wave are not actually simple positive or negative numbers, but complex numbers. These can be imagined as arrows or clock hands in a two-dimensional plane (see Fig. 1.4).

But what is such a strange electron wave supposed to mean physically? We already know that wavelength and wave frequency determine the energy and

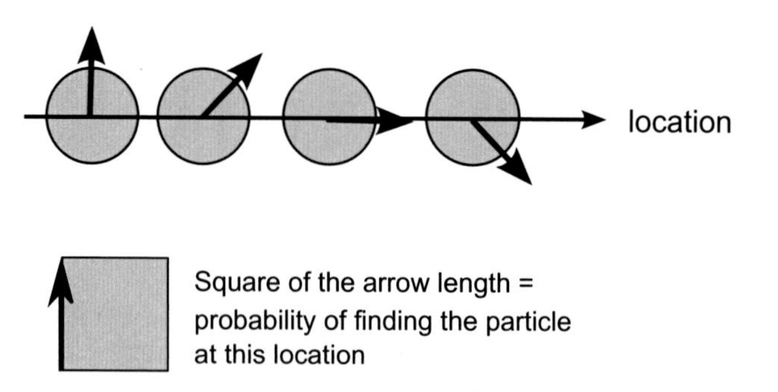

Fig. 1.4 Snapshot of a quantum wave. The values of the wave function at individual locations can be pictured as given by rotating arrows or clock hands

momentum of the particle. But where is the particle? A wave is a spatially extended object, while a particle should always be in a certain place!

The solution to this problem is both ingenious and strange: it is precisely this demand that is dropped, that a particle always stays in a certain place and that it moves on a certain path. Instead, one goes to a description by probabilities, where the square $|\psi|^2$ of the wave function – i.e., the squared height of the quantum wave or, more precisely, its squared arrow length – indicates the probability of finding the particle at the corresponding location. In other words, as long as the location of the particle does not leave any detectable traces in our world, it is basically indefinite. In a way, the particle itself does not know where it is. Only when a sufficiently sensitive interaction with the environment (or a measuring device) takes place does it make sense to speak of a particle location. It is exactly then that the probability interpretation of the wave function takes effect.

We may of course ask why we need a quantum wave at all. Why do we not work with probabilities right away? The reason for this is that the wave crests and troughs of a quantum wave can erase each other when they meet, while two probabilities always add up to a greater overall probability. This phenomenon, known as interference, will be of great importance to our discussion later on.

Why is it that in quantum mechanics the concept of probability suddenly comes into play? We don't actually know. This is one of the greatest mysteries: why do we have to deal with probabilities in a theory that we still consider fundamental today? Randomness seems to play a fundamental role in nature. If you do not want to believe this, you are in good company, because Albert Einstein also had his doubts, expressing them in his famous remark: "God does not throw dice". But for all we know today, God does indeed seem to throw dice. In his 1979 lecture on quantum electrodynamics (QED) at the University of Auckland (New Zealand), Feynman expressed this idea in the following way[4]:

> If you want to know how nature works, we looked at it, carefully. Looking at it, that's the way it looks. You don't like it? Go somewhere else, to another universe where the rules are simpler, philosophically more pleasing, more psychologically easy. I can't help it, okay?

This is typical of Feynman. It is not our desires that are important, but only reality – no matter what we think of it. And yet the fundamental

[4]A very entertaining video can be found, for example, at https://www.youtube.com/watch?v=iMDTcMD6pOw

role of randomness in quantum mechanics remains strange and is still not well understood today. Feynman also admits this when he says elsewhere: "Nobody understands quantum mechanics!"

In 1926, the Austrian physicist Erwin Schrödinger formulated the basic quantum mechanical wave equation, a differential equation that can be used to calculate the evolution of wave functions. And it turned out that the results obtained with the Schrödinger equation were in perfect agreement with experiment, so physicists knew they had found the right approach to the atomic world.

The theory of quantum mechanics founded by Louis de Broglie, Niels Bohr, Erwin Schrödinger, Werner Heisenberg, and many others was to develop quickly (see Fig. 1.5). In the end, a fundamental theory was built up, describing nature on the microscopic level. The structure and properties of atoms, molecules, solids, and much more could be calculated, at least in principle. The Special Theory of Relativity was also included shortly afterwards, when the British physicist Paul Dirac established his famous Dirac equation in 1928. It was also Paul Dirac who published the first comprehensive textbook on quantum mechanics in 1930: *The Principles of Quantum Mechanics*. And using this, Feynman and his fellow students became the first generation of physicists to be introduced to the new quantum mechanics as part of their studies.

Fig. 1.5 Participants at the Solvay Conference on Quantum Mechanics 1927. Photograph by Benjamin Couprie, Institute International de Physique Solvay, Brussels, Belgium

Feynman, the Free-Thinker

As in high school, Feynman learned a lot from books and from working with other talented students at MIT. He was interested not only in physics, but also the other natural sciences, such as chemistry and metallurgy, and passed all his science exams with very good results. One problem, however, was the humanities subjects and languages, of which he had to take three. Fortunately, astronomy was one of them – that was fine with him. English was compulsory, and as the third subject he chose philosophy, which he didn't like at all. The lectures at MIT only deepened his natural aversion and strengthened his conviction that philosophy was just meaningless verbiage. Later, as an established physics professor, he would gladly use every opportunity to knock the philosophers.

Feynman's intellectual independence was also illustrated by his refusal to solve mechanical exercises using the Lagrangian method (see Infobox 1.2). He insisted on applying the original Newtonian law of motion and splitting all the forces into different proportions to suit the problem at hand. The Lagrangian method already incorporates this maneuver, so the solution can be approached schematically. But that was obviously not interesting enough for Feynman. Perhaps he was not aware at that time that the Lagrangian method was closely related to the principle of least action that had fascinated him so much at high school – otherwise he might have shown greater interest in this elegant method. But instead, he preferred to train his physical intuition rather than calculating in this schematic way. He would ponder over a task in a highly concentrated manner and try to illuminate it from various angles until he had the solution. Indeed, he would often guess the solution and only check its correctness afterwards. These are exactly the qualities that a great physicist should have, and that the young Albert Einstein, for example, also possessed.

Infobox 1.2: Lagrangian Method

In 1687, Isaac Newton established his famous law of motion, which says that *force equals mass times acceleration* or $F = m \cdot a$ in the obvious notation, and thus laid the foundations of mechanics. He and Gottfried Wilhelm Leibniz also independently invented the infinitesimal calculus, i.e., the differentiation and integration of functions, and thereby created a powerful mathematical tool that led to rapid developments in mathematics.

To be able to apply Newton's law of motion, Cartesian coordinates x, y, and z are usually used in three-dimensional space. However, these orthogonal coordinates are often not well adapted to the physical problem, resulting in quite complicated equations. For example, for the orbit of a planet around the Sun, it is much easier to use its distance from the Sun and an angle variable indicating its position on the orbit. Instead of working with x, y, and z, it's better to work with distances and angles, which are examples of generalized coordinates. However, some effort has to be put into express Newton's law of motion for a planet in terms of distances and angles – and that was exactly the way Feynman loved to work.

About 100 years after Newton, the French mathematician Joseph-Louis Lagrange succeeded in finding a general method for establishing equations of motion in arbitrary coordinates – the so-called Lagrangian method. For example, when a planet moves around the Sun, this method provides the equations of motion directly in the desired form, i.e., expressed in terms of the planet's distance from the Sun and its angle variable.

In principle, the method works as follows:

1. We introduce time-dependent coordinates and associated velocities that are adapted to the problem (e.g., distance and angle variables together with their rates of change).
2. We now express the kinetic energy T and the potential energy V in terms of these coordinates and velocities and form the difference $L = T - V$, which is called the *Lagrangian*.
3. The equations of motion for each of the coordinates are now obtained in this way:

- For each individual coordinate, we form the derivative of the Lagrangian L with respect to the coordinate velocity and then with respect to time t.
- We then set the result equal to the derivative of L with respect to the corresponding coordinate.

If you feel your mathematics is up to it, you can test this procedure yourself using the simple example of the spring pendulum. But first of all, let's proceed in the conventional way. In such a pendulum, a mass m hangs on a spring with spring constant D and is set into vertical oscillation (see Fig. 1.6). The spring pulls the mass back in the direction of the resting position with the force $F = -D \cdot x$, where x is the distance from the resting position (so that $x = 0$ in that position). If we use this force in Newton's law of motion, we obtain the equation of motion

$$m \cdot a = -D \cdot x$$

and now consider the same situation using the Lagrangian method. With the kinetic energy $T = m \cdot v^2/2$, where v is the velocity, and the potential energy $V = D \cdot x^2/2$, we get the Lagrangian

$$L = \frac{m}{2} \cdot v^2 - \frac{D}{2} \cdot x^2.$$

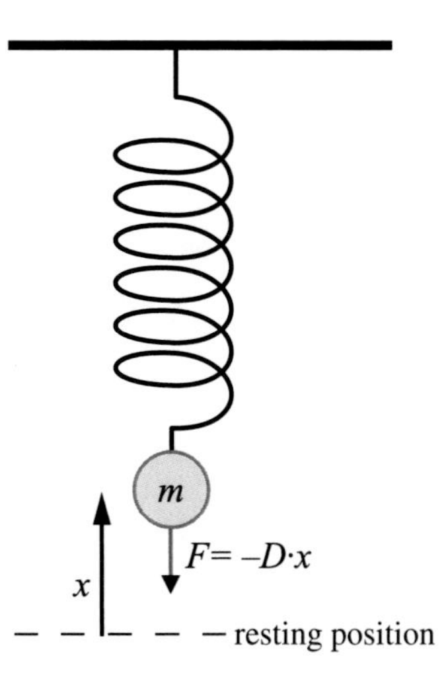

Fig. 1.6 The spring pendulum

Differentiating the Lagrangian L with respect to v, we obtain $m \cdot v$, and then differentiating with respect to time gives $m \cdot a$, because the acceleration a is the derivative of the velocity v with respect to time. This is the left side of the equation of motion. Differentiating L with respect to the coordinate x results in the term $-D \cdot x$ and we have the right side of this equation.

With the spring pendulum the advantages of the Lagrangian method are not yet really evident, since no special coordinates are needed. This is different for the motion of the planets around the Sun – here the Lagrangian method is much easier than the direct calculation of forces using a distance and an angle coordinate.

The procedure can also be generalized to fields, and ultimately even to all physical laws known today. If you know the Lagrangian, then you know the physics – at least in principle.

Theory of Relativity and Quantum Waves

Of course, Feynman was not the only highly talented physics student at MIT. He especially liked his fellow student Theodore Welton, also known as Ted. Both were intelligent and ambitious, and together they tried to go more deeply into the secrets of physics. Feynman loved to deduce everything as far as possible in his own way, and Ted Welton was an ideal companion in this venture.

One day, Feynman and Welton considered how the quantum mechanical Schrödinger equation could be reconciled with Einstein's Special Theory of Relativity. They had already dealt intensively with Special Relativity – so let's take a closer look at the core idea of this cornerstone of modern physics.

When Isaac Newton formulated his laws of mechanics at the end of the seventeenth century, he envisioned an absolute space in which an object either rests or moves. Even today, this idea still corresponds to the usual human view of things.

But consider, for example, some astronauts in a windowless space capsule floating through empty space with its propulsion switched off. They will not feel any motion of the space capsule. No mechanical experiment will allow them to determine whether the capsule is stationary or moving. Newton was already aware of that, too. As early as 1632, Galileo Galilei argued that in a room below the deck of a large ship, one would not notice whether the ship was moving steadily or standing still. In this way he could explain why we do not feel the motion of the Earth on its way around the Sun.

Can motion in absolute space be detected in some way using a ray of light? Imagine that you are an astronaut in a space capsule at rest, and you measure the speed of a short burst of light propagating from the stern towards the bow. The value obtained will be 300,000 km/s.

If the space capsule now moves in the bow direction, you can follow the light pulse that you sent out in the bow direction. If light always moves at 300,000 km/s in absolute space and you follow it at the crazy speed of 100,000 km/s, then the light pulse should move at 200,000 km/s from your point of view. Inside the racing space capsule you should notice a reduced speed of light in the bow direction. Conversely, you can conclude from this speed reduction that your space capsule must be moving at 100,000 km/s towards the bow.

It all sounds perfectly logical. Now, many attempts were made to measure this effect, but they found – NOTHING! No matter how fast the space capsule moves – light always moves inside it at 300,000 km/s. As an astronaut in the capsule, you would not notice any reduction in the speed of light in the direction of flight.

It looked as if nature had conspired against the experimenters and invented all sorts of tricks to make the effect invisible. Albert Einstein first solved the problem by making this apparent conspiracy the basic principle of his theory. With regard to our space capsule, this means:

> In a space capsule that glides through space with the engines turned off, no physical experiment within the capsule can be used to find out whether this

capsule is at rest or moving uniformly without acceleration. This applies not only to mechanical experiments, but also to light. From the astronauts' point of view, the speed of light always has the same value c in every direction inside the space capsule.

So you can't catch up or overtake light and you can't push it either. No matter how fast the space capsule moves – from the astronaut's point of view within the capsule, every ray of light always moves at the speed of light. This also means that the space capsule and any other massive object can never go faster than light.

This basic principle discovered by Einstein contradicts our intuition, as we always imagine an absolute space and an absolute time, like Newton. It is precisely these concepts that we have to give up in order to understand the strange constancy of the speed of light. And it leads to strange conclusions, all of which were eventually confirmed by experiment – here is an example.

If two identical space capsules carrying identical clocks pass each other very quickly, an astronaut in capsule 1 will find that capsule 2 is shorter in the direction of motion than his own capsule, and that the clock on board capsule 2 ticks more slowly than his own clock. These effects are called *length contraction* and *time dilation*. Furthermore, an astronaut in capsule 2 would say the same thing about capsule 1 – and both would be right!

In Special Relativity, observers moving relative to each other will assess spatial and temporal distances differently. The mathematical formulas used to formulate the laws of nature must take this into account. In mechanics, this leads to the following relation between the energy E, the mass m, and the momentum p of an object:

$$E^2 = \left(m \cdot c^2\right)^2 + (p \cdot c)^2.$$

Here c is the speed of light. For an object at rest (i.e., with zero momentum), we get the most famous formula of the world: $E = m \cdot c^2$, which tells us that, apart from the conversion factor c^2, mass is nothing other than energy trapped in the resting object.

The above relationship between energy, mass, and momentum is a starting point to search for relativistic wave equations that can be used to describe quantum waves – for example, for the electrons in an atom. When Feynman and Welton tried to do this, they came across the *Klein–Gordon equation*, which was already well-known at the time (see Infobox 1.3). With this equation, they hoped to be able to calculate the energy levels of the hydrogen atom even more precisely than with the non-relativistic Schrödinger equation – at least that is what they were hoping.

However, the result was rather unsatisfactory and did not agree with experimental results, even though their approach was perfectly cogent, since the Klein–Gordon equation is an obvious relativistic generalization of the Schrödinger equation. However, there is another relativistic equation that is not nearly so obvious, formulated by Paul Dirac a few years earlier: the *Dirac equation*. This is the correct equation for the electron in the hydrogen atom, taking into account its spin. The electron has spin 1/2, i.e., it carries a certain quantum mechanical angular momentum, as if it were rotating about its own axis. By contrast, the Klein–Gordon equation describes particles without spin. This experience served as a warning to Feynman, showing just how quickly one could be mistaken, and he grew to mistrust every theory, no matter how beautiful, until it had proved its worth in experiment.

Infobox 1.3: Quantum mechanical wave equations

Are you already familiar with physics and mathematics? Would you like to see the Schrödinger equation for real? And would you like to see what the Klein–Gordon and Dirac equations are all about? Then this infobox may be the right thing for you – otherwise you can simply skip it.

To guess the quantum mechanical wave equation of a particle, there is a simple recipe: you start with the relationship between the energy E and the momentum p. We already know that the energy of a particle has something to do with the frequency of the corresponding quantum wave, i.e., how fast the wave oscillates and changes over time. Similarly, the momentum has something to do with the wavelength of the wave, i.e., how quickly it changes in space. Mathematically, the rate of change of the wave ψ with time is given by the derivative $d\psi/dt$, while its rate of change in space is expressed by the spatial derivative $d\psi/dx$. If squares appear in the relationship between the energy and the momentum, we have to use the second derivatives in time or in space.

All this can be deduced in a mathematically precise way, something we don't wish to enter into here. But such an analysis informs us that suitable pre-factors such as $i \cdot \hbar$ should be included, where i is the imaginary unit of the complex numbers, with $i^2 = -1$, and $\hbar = h/(2\pi)$ is the Planck constant h divided by 2π. We encountered complex numbers a little earlier in our discussion. The values of the quantum wave ψ are just such complex numbers, which can be visualized as arrows or clock hands in a two-dimensional plane, as in Fig. 1.4. The imaginary unit i corresponds to an upward pointing arrow of unit length.

For a particle with mass m which moves much more slowly than the speed of light c, the non-relativistic relationship $E = p^2/(2m) + V$ applies. The term containing the square of the momentum is the kinetic energy, while V is the potential energy. The corresponding wave equation is the Schrödinger equation

$$\left(i\hbar\frac{d}{dt}\right)\psi = \left(-\frac{\hbar^2}{2m}\frac{d^2}{dx^2} + V\right)\psi.$$

This describes how the quantum wave ψ evolves in time and how it oscillates through space. Where we previously had the energy E, we now see the time derivative of the wave (with a pre-factor), and where we previously had the square of the momentum, we now see the second spatial derivative of the wave (also with a pre-factor).

At higher velocities, which may also come close to the speed of light c, the relativistic equation $E^2 = (p \cdot c)^2 + (m \cdot c^2)^2$ applies, where we have omitted the potential energy V for simplicity. Replacing energy and momentum by the temporal and spatial derivatives of the quantum wave, then including the appropriate pre-factors, leads to the Klein–Gordon equation. The squares E^2 and p^2 become second derivatives with respect to time and space, respectively, while the Schrödinger equation contains only the first derivative with respect to time.

The second time derivative in the Klein–Gordon equation leads to certain mathematical complications, and this was why Dirac sought a relativistic equation, which contained only the first time derivative, like the Schrödinger equation. He found this equation in 1928 using the following trick. He first wrote the relativistic energy–momentum relationship in the form $E = \alpha \cdot (p \cdot c) + \beta \cdot (m \cdot c^2)$, where E and p are not squared, so that no second derivatives will occur. In order to ensure that we obtain the correct relativistic energy–momentum relationship when squaring the equation, we must have $\alpha^2 = \beta^2 = 1$ and $\alpha \cdot \beta + \beta \cdot \alpha = 0$. However, this is not possible with numbers. It turns out that α and β must be matrices. In three space dimensions, two additional α-matrices are needed for the other momentum components, and this means that four matrices are required to meet all the conditions.

In this energy–momentum relationship, the energy and momentum can now be replaced by the corresponding derivatives, and the Dirac equation is obtained.

During his studies at MIT, Feynman discovered a new passion that had nothing to do with physics: he began to play bongos. This was a hobby he would keep up for the rest of his life, always perfecting it. Although he claimed not to be particularly knowledgeable about music, he had rhythm in his blood.

He was an attractive man, 1.80 m tall, and certainly had plenty of opportunities with the girls, but Arline remained his great love. They wrote and met as often as possible, especially during the semester break. Eventually they decided to marry as soon as Feynman had finished his studies, something which caused mixed feelings among Feynman's parents. They liked

Arline, but Melville was afraid that this might have a negative effect on their son's studies, and he didn't want this to jeopardize his chances. Lucille was also worried – how could a poor student support a family? A conflict arose between Feynman and his parents, which would deepen later on. Melville and Lucille underestimated their son's deep affinity with Arline and the importance she played in his life.

Forces and Stresses in Molecules

In his fourth and final year at MIT, Feynman wrote his bachelor's thesis, which consisted of thirty typewritten pages with the mathematical formulas inserted by hand. It clearly demonstrates his outstanding skills and is a pleasure to read, with Feynman's unique talent for explaining physical facts coming vividly to the fore. And so we find that Feynman was already a great teacher at the age of 20!

The thesis is entitled *Forces and Stresses in Molecules*, and was later published in a slightly modified form in the prestigious journal Physical Review (just like a somewhat earlier work on cosmic radiation, written with Manuel Sandoval Vallarta, which we shall not discuss here). The results of Feynman's bachelor thesis are still relevant today and are known as the Hellmann–Feynman theorem, since the German physicist Hans Hellmann also made important contributions to it.

What did Feynman (and Hellmann) discover? They found a simple method for calculating the forces between the atoms in a chemical molecule. If one knows these forces, then one can understand the spatial arrangement of the atoms in the molecule. This works essentially as follows. Molecules have to be described using quantum mechanics. In order to solve the complicated Schrödinger equation to a good enough approximation, one can assume that the centers of the atoms, i.e., the heavy atomic nuclei, are located at certain points in space. One then calculates the quantum wave of the much lighter electrons in the electric field of the nuclei. This gives the total energy of the molecule for the given positions of the nuclei. Forces do not come into this description, because there is no concept of force in the quantum mechanical equations.

One can now change the positions of the nuclei in space and calculate the total energy of the molecule for the new positions. In the simple case of a hydrogen molecule, we would thus consider the energy of the molecule for different distances between the two hydrogen nuclei. If we then plot a graph with the distance on the x-axis and the total energy on the y-axis, we get a simple curve with a minimum energy at a certain distance (see Fig. 1.7).

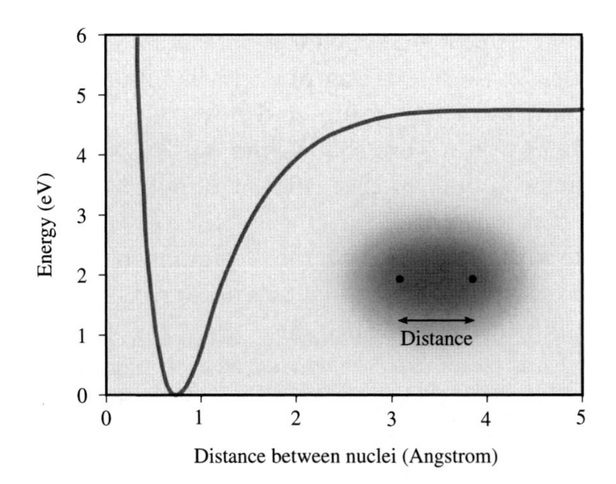

Fig. 1.7 Energy of the hydrogen molecule for different distances between the two nuclei. The minimal energy is at a distance of about 0.74 Å, where one angstrom corresponds to a ten-millionth of a millimeter

This is the equilibrium distance that one would expect to find between the nuclei in a hydrogen molecule.

If the distance between the nuclei is greater or smaller than the equilibrium distance, a force will act on the nuclei, driving them toward the minimal energy. This force is given by the slope of the energy curve in the graph and directed downhill – the molecule tries to move down the curve until it reaches the lowest energy. The gradient of the energy curve can then be calculated to determine the force between the two hydrogen nuclei.

What Feynman showed in his bachelor's thesis was that there is an easier way: the force can be calculated directly from the mean charge distribution of the electrons between the nuclei, and this in turn can be determined by solving the Schrödinger equation. On a statistical average, the electrical forces between the electrons and the atomic nuclei then constitute the force acting between the two hydrogen atoms in the molecule. This is just the way chemists like to imagine the binding mechanism in molecules, and rightly so, as Feynman showed.

The above result may seem trivial to some, but one must first prove that the classical consideration of forces between charges does indeed provide the same result as the energy curve from the quantum mechanical calculation. This was what Feynman succeeded in doing.

Feynman completed his studies at MIT with his remarkable thesis in 1939. He would have liked to have stayed there for his doctoral thesis, but John Slater, who was head of the Department of Physics at MIT at that time and who had helped to supervise Feynman's work, would not allow him to do so. For a promising young physicist like Feynman, it was important to get to know other universities and broaden his horizons. Feynman later recalled Slater's words: "You should find out how the rest of the world is," and was grateful to him for that later on, although in the immediate it meant finding a new university.[5]

In contrast to what happened at the beginning of his studies, he no longer had to struggle to be accepted at a university. He had written an excellent bachelor's thesis and had also won the prestigious *William Lowell Putnam Mathematical Competition*, which earned him a scholarship at nearby Harvard University. However, Feynman chose Princeton, which was another very prestigious university located southwest of New York. The famous Albert Einstein had also found a new home there, having fled from the Nazis in 1933. Feynman's relatively poor grades in history and English and his Jewish background might almost have been fatal, but his outstanding achievements in physics and mathematics, combined with Slater's personal influence, were deciding factors and he was duly admitted.

Before we go there, however, let's take a closer look at two closely related physical principles that would eventually have a decisive influence on Feynman's future scientific career: Fermat's principle and the principle of least action, which we already outlined briefly above.

1.2 Light Saves as Much Time as It Can: Fermat's Principle

When Feynman's high school teacher Abram Bader introduced his student to the principle of least action, he could have told him something about another principle. Perhaps he did actually do so – at least it would have been a good idea, because the two principles are closely related. The principle in question is Fermat's, which can be viewed as a special case of the principle of least action. This principle tells us which path light rays take when they hit a mirror, for example, or when they go from one medium such as air into another such as water. Later, this principle will help us to understand Feynman's approach to quantum mechanics.

[5]See *Surely You're Joking, Mr. Feynman.*

Reflection and Refraction of Light

The law that determines the reflection of a light beam on a mirror is very simple: the angle which the incident ray makes with the normal is equal to the angle which the reflected ray makes with the same normal, but on the opposite side of that normal (see Fig. 1.8).

When it is reflected by a mirror, a beam of light changes its path in a very simple way. But light can also change its direction for other reasons, such as when the Sun's light enters the clear water of a calm lake. The law governing the way it changes direction when passing from a medium like air into another medium like water is somewhat more complicated than the law of reflection. It goes by the name of *Snell's law of refraction*, its discovery being attributed to the Dutch astronomer and mathematician Willebrord van Roijen Snell in 1621. The law states that the angles α_1 and α_2 of incidence and refraction relative to the normal of the boundary satisfy the following relation:

$$\frac{\sin\alpha_1}{\sin\alpha_2} = \frac{n_2}{n_1}.$$

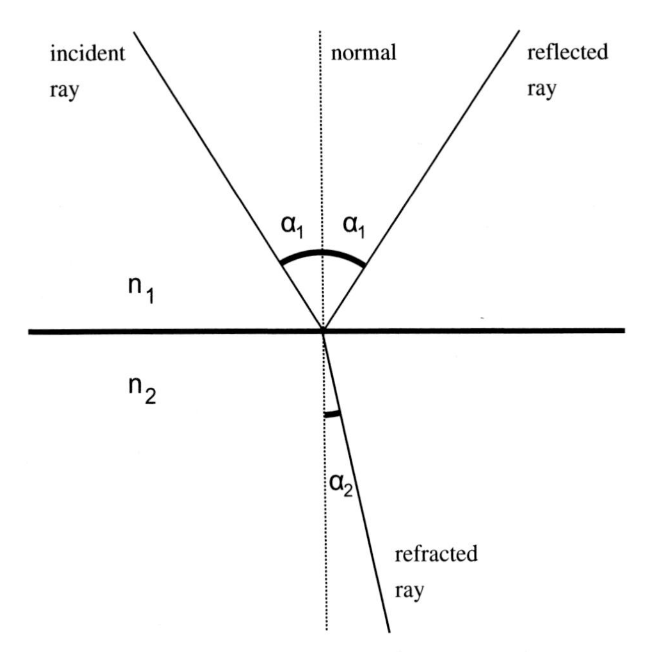

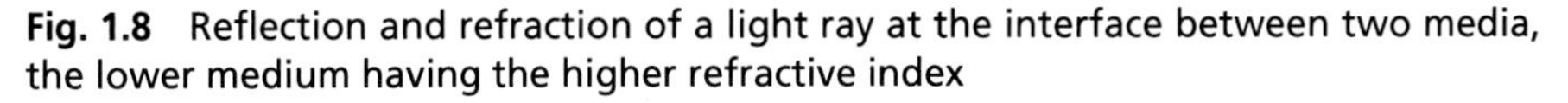
Fig. 1.8 Reflection and refraction of a light ray at the interface between two media, the lower medium having the higher refractive index

The symbols n_1 and n_2 denote the refractive indices of the two media. They are positive numbers that express the refractive power of the corresponding medium, the refractive index 1 being assigned to the vacuum. Air has a refractive index very close to 1, water stands at 1.33, and diamond at 2.42. The law states that the light beam is refracted towards the normal when going from the optically thinner to the optically denser medium – i.e., from the lower to the higher refractive index. The refracted beam is then more perpendicular to the boundary surface between the media than the incident one (see Fig. 1.8). This can be illustrated by a simple experiment: use a laser pointer to shine diagonally into a glass of water from above and follow the path the light beam takes (this works especially well with a cup of tea).

What the empirical laws of reflection and refraction do not tell us is why light behaves in this way. For example, we do not know where the refractive index of a medium comes from, only that it is needed to express the relation between the two angles in the law of refraction.

Is there perhaps a higher principle from which both laws can be derived? There is indeed! A first attempt to find such a law was in fact made around 2000 years ago. In the 1st century A. D., the Greek mathematician and engineer Heron von Alexandria already suspected that light simply takes the shortest path from the starting point to the mirror and then to the target point. He was thus able to explain the law of reflection. Figure 1.9 shows that the path obeying the law of reflection is indeed the shortest path between points A and B with mirror contact. The path is exactly as long as the path that runs straight through the mirror to point B′. If we move the

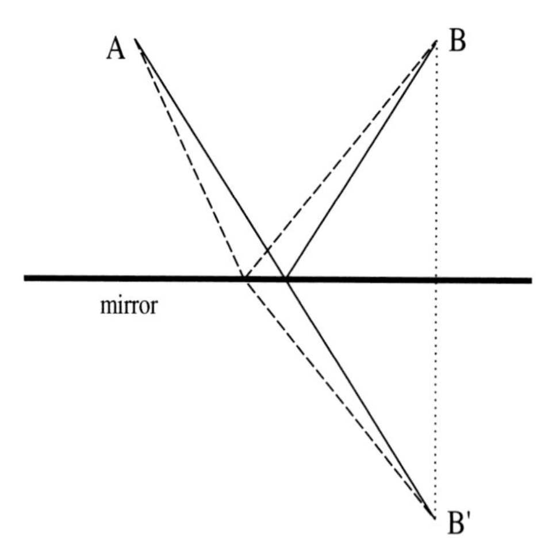

Fig. 1.9 When reflected from a mirror, light takes the shortest path between points A and B with mirror contact (solid line)

contact point on the mirror surface, the distance to B′ is increased because the straight path gets a kink (dashed line in the graphic), and the equally long distance from the new mirror contact point to B is therefore also increased.

However, the law of refraction cannot be explained using Heron's idea, because the refracted light beam does not take the shortest route between the start and end points – that would simply be the straight path without any refraction at all. But perhaps the idea can be modified in such a way that we can continue to explain reflection, but also cover refraction.

Fermat's Principle and the Speed of Light

Around the year 1650, the French mathematician and lawyer Pierre de Fermat had the decisive idea. He formulated the following law, which is also known as Fermat's principle:

> Light does not necessarily choose the shortest route between two points, but the fastest route – i.e., the one it can travel in the shortest possible time.

In order to apply this law, we must of course say something about the speed of light. In Fermat's day, the speed of light was still a subject of intense discussion. For example, Fermat's contemporary René Descartes was of the opinion that light must propagate infinitely fast, because all previous measurements had been unable to identify a finite speed. It was not until 1676 that the Danish astronomer Ole Rømer succeeded for the first time in demonstrating the finiteness of the speed of light by observing the motion of Jupiter's moon Io through his telescope. He noticed that Jupiter's moon always entered the shadow of Jupiter during its orbit somewhat later than calculated, when the Earth was further away from Jupiter and Io. Half a year later, when Earth and Jupiter were closer again, Io entered the shadow of Jupiter somewhat earlier. Rømer concluded that the light could not be travelling infinitely fast from Io to Earth – it was taking more time to reach Earth when the distance was greater, and this was why he saw Io entering the shadow of Jupiter later.

Unfortunately, Rømer was unable to determine the speed of light, because astronomical distances were so poorly established at that time. Later, the Dutch scientist Christiaan Huygens calculated a value of about 200,000 km/s on the basis of Rømer's observations and those of other astronomers. This was already of the right order of magnitude.

It was only in 1728 that the English clergyman and astronomer James Bradley was able to determine the correct value of the speed of light in a vacuum – almost 300,000 km/s – with an accuracy of up to one percent. In air, the speed of light is only slightly lower.

The law of reflection is directly compatible with Fermat's principle, because at a constant speed of light, the fastest way is also the shortest way. In order to include the law of refraction, Fermat had to assume that in a medium with refractive index n the light is slower than in a vacuum by a factor of $1/n$. In water, with $n = 1.33$, light is therefore 1/1.33 or 0.75 times slower. If a ray of light now wants to get from a point in the first medium to another point in the second medium as quickly as possible, it will not choose the straight path, but will instead introduce a kink along its way.

The path in the faster medium is thus longer, while the path in the slower medium is shorter, which saves time overall. A lifeguard who wants to reach a victim floating in the sea as quickly as possible uses the same method: he will run as far as possible along the beach to shorten the more laborious part of the path through the water. If you calculate this exactly, you will find out that the fastest way is exactly the one that satisfies the law of refraction.

But do we gain anything from this new perspective? The laws of reflection and refraction also describe the behavior of light correctly – so why do we need Fermat's principle?

First of all, an overarching law is certainly more satisfactory than two separate and independent laws. But Fermat's principle does more: it provides a deeper reason for the behavior of light in two different situations. From this, further predictions can be derived which go beyond reflection and refraction. Fermat's principle predicts in particular that light has different speeds in different media such as air, water, or glass. And this can be checked experimentally! We can measure the speed of light in these media and see if it really is slowed down by the factor $1/n$ in relation to the vacuum, where n is the refractive index of the medium, something that can be determined from the angle of refraction. And indeed, Fermat's principle passes this test with flying colors! Unfortunately, Fermat himself was unable to witness this, for he died in 1665, eleven years before Ole Rømer succeeded in demonstrating the finiteness of the speed of light for the first time.

Apparently, with the help of Fermat's principle we can understand light on a more fundamental level than is possible with the laws of reflection and refraction alone. So light chooses the fastest path between two points for some reason. But why does it do that? Is there an explanation for this, or do we simply have to accept Fermat's principle as God-given without any further explanation?

It turns out that there is indeed a deeper reason, because the description of light by light rays that we have used so far is only an approximation that does not fully grasp the true nature of light. If we take a closer look, we see that light is an electromagnetic wave, and we shall explore its properties in the next section.

Waves Take Every Possible Path

Two properties of waves lead to Fermat's principle: the ability for constructive and destructive interference and the possibility of deviating from the straight path followed by light rays.

Interference means that two or more waves superpose in such a way that individual wave crests and troughs either add or subtract to form a new resultant wave. Where a crest of one wave meets a crest of the other wave, the amplitudes add up (constructive interference). However, if a wave crest meets a wave trough, the waves tend to cancel each other (destructive interference). This is the key feature of interference: wave plus wave can produce nothing if the phases of the waves – i.e., the relative positions of wave crests and troughs – are set in such a way that crests meet troughs.

The second characteristic – that waves can deviate from the straight path of classical light rays – leads to the phenomenon of diffraction. For example, waves can bypass obstacles if their wavelength is similar to the size of the obstacle. With sound waves, diffraction is easy to observe due to their relatively long wavelengths. It means that sound can reach areas that could not be reached by moving along a straight line. That's why we can hear sound sources without seeing them.

Diffraction is easily observed when a wave passes through a narrow slit. If the slit is much wider than the wavelength, then the wave will pass straight through like a beam – this is exactly the situation with light when it shines through a not too small pinhole aperture and forms a sharply limited light beam behind it. If the slit is now reduced, the beam will become increasingly blurred, and part of the wave will also reach other areas (see Fig. 1.10). When the slit is only about as big as the wavelength, semi-circular waves will form behind the slit – the wave now takes all the paths that lead from the slit into the region behind it.

Huygens' principle, discovered in 1678 by the Dutch physicist Christiaan Huygens (thirteen years after Fermat's death), provides a good explanation of this behavior. Huygens' idea was that each point of a wave front would become like a source for a small spherical wave (called a *wavelet* or *elementary wave*), propagating at the same speed as the original wave. All these

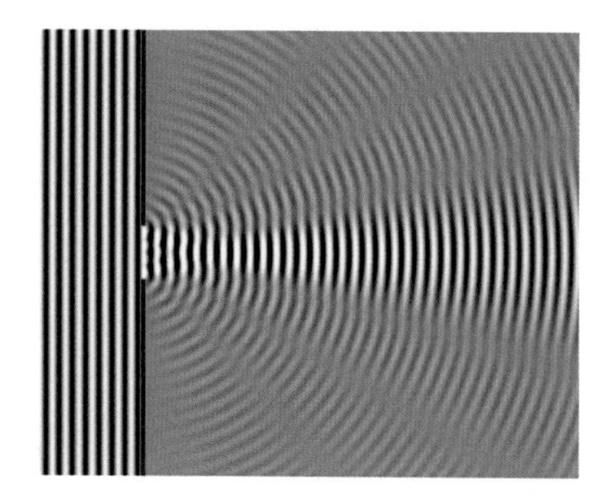

Fig. 1.10 Diffraction pattern from a slit with width equal to four wavelengths. Source: https://commons.wikimedia.org/wiki/File:Wave_Diffraction_4Lambda_Slit.png, public domain

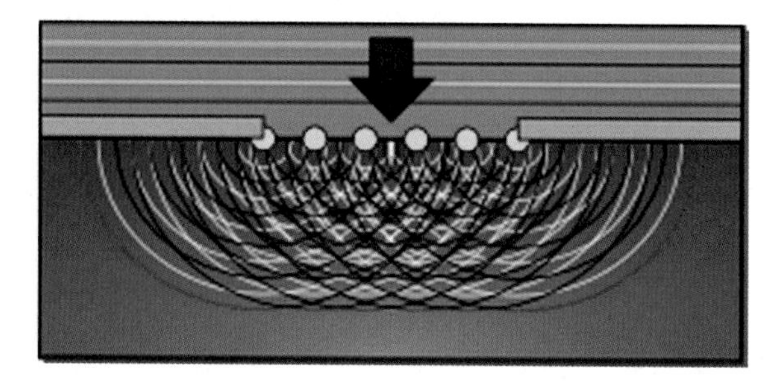

Fig. 1.11 Diffraction at a slit according to Huygens' principle. Six elementary waves are shown, their origins marked by yellow circles. Wave crests are white, wave troughs black. Source: https://commons.wikimedia.org/wiki/File:HuygensDiffraction.svg, public domain

elementary waves then add up to form a new wave, i.e., they interfere constructively or destructively with each other. If we look at a very long wave front, a new long wave front is created in this way by interference of all the elementary waves, and this is effectively what produces the propagation of the wave at any subsequent time.

If a plane wave now passes through a slit, its wave front will be cut off at either side. The elementary waves emanating from the edge areas of the slit then lack their interference partners from the cut-off areas of the wave front, so they can penetrate relatively undisturbed into the lateral areas behind the slit and the wave is diffracted (see Fig. 1.11).

The interplay of elementary waves can also be interpreted in a slightly different way. In the above, we only picked out a few elementary waves.

However, we can also consider the spherical wave front of each elementary wave as itself a source for many new elementary waves, which then generate further elementary waves, and so on. A field of tiny elementary wavelets is created, which all overlap to form the overall wave. If we now pick a point from any elementary wave and connect it to a point on the next elementary wave created there and so on, we can construct any path taken by the waves – we only have to select the elementary waves small enough and connect suitable points on the wave fronts with each other (Fig. 1.12). *In this sense we can say that the wave takes every possible path.*

This idea has the following consequence: instead of talking about all conceivable elementary waves, we can instead think about the wave contributions of all possible paths – the result is the same. We can imagine that a tiny wave runs along every path.

Let's take a look at the reflection of light from a mirror (Fig. 1.9). What is the contribution of the individual paths from point A to a point on the mirror and then to point B?

We can move the point on the mirror back and forth and thus systematically grasp all paths. There is exactly one fastest and also shortest way: the one with the angle of the incident ray equal to angle of the reflected ray, as we already know. As soon as we move the contact point to the right or left, the path length gets longer – at first very slowly and then ever faster.

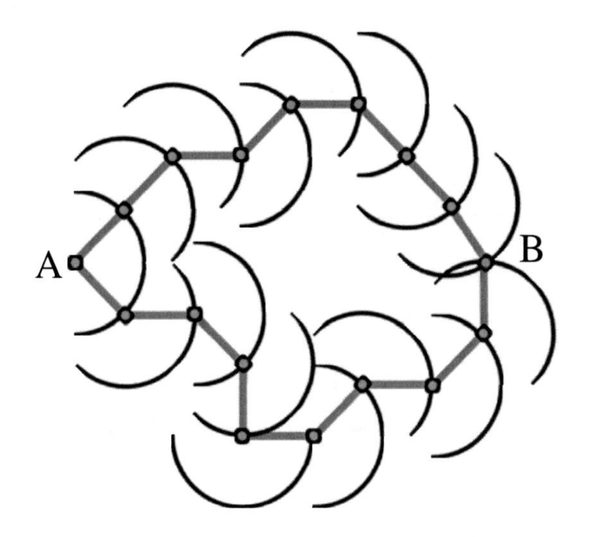

Fig. 1.12 The constantly developing elementary waves ensure that a wave takes all conceivable paths at the same time. The figure shows two example paths from point A to point B

The paths in the immediate vicinity of the shortest path thus have almost the same length as the shortest path, while at a greater distance the path length changes ever faster (Fig. 1.13).

Depending on the length of the path, a different number of wave crests and troughs fit along it, and the wave amplitude at point B can have different values for each path. Further away from the shortest path, the lengths of neighboring paths change quickly, so the wave amplitudes of the individual paths vary quickly and cancel each other to a large extent at point B.

Close to the shortest path, the situation is different: since the lengths of the paths do not change very much there, we have many paths with approximately the same wave amplitude at point B as the shortest path itself. All paths whose path lengths differ by less than half a wavelength from the shortest path make a contribution in the same direction, so the waves of these paths interfere constructively with each other. The light therefore selects the shortest path and nearby paths, while the other paths hardly play a role. If we now look at this situation from a distance, the shortest path

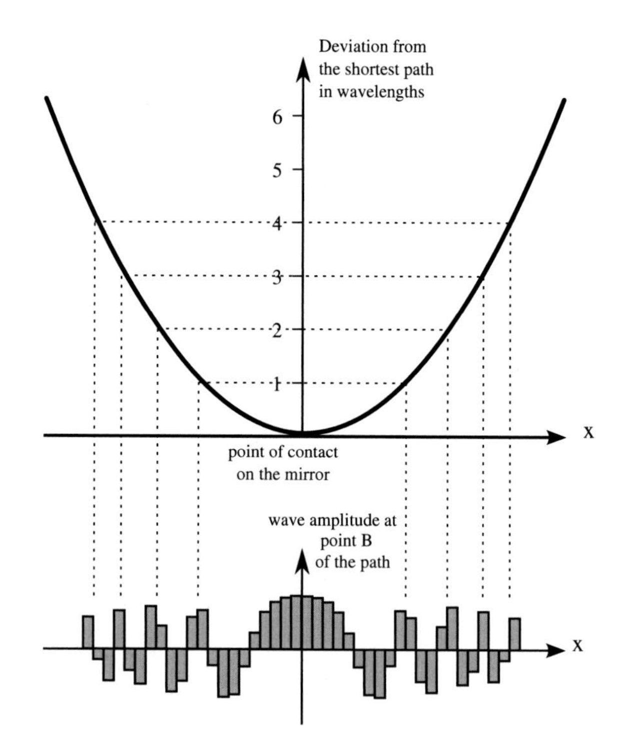

Fig. 1.13 Lengths of the different paths and associated wave amplitude at the endpoint B for light reflection by a mirror, provided that the shortest path has a wave crest at the endpoint B

seems to merge with its nearby paths and it looks as if a single thin beam of light takes the shortest path.

This is exactly Fermat's principle for the reflection of light rays, as discussed above. With the help of the wave description, we have succeeded in finding a justification for this principle. The Fermat principle is not therefore a divine law, but an approximation that will apply to any wave! It is always applicable when many neighboring paths can be taken whose lengths differ by less than half a wavelength. The dimensions of the entire arrangement must be significantly greater than the wavelength, so that, on the scale of the setup as a whole, the individual waves are virtually invisible. Since light has a wavelength between 400 and 800 nm (billionths of a meter) depending on the color, this condition is usually satisfied on the length scales that concern us in everyday life. The details of light waves then no longer play a role and we can use light rays as a good approximation, together with Fermat's principle, to describe the behavior of light.

In the general case of light refraction between different media, the same reasoning applies, except that we must consider here the fastest and not the shortest way. Where the light moves more slowly, the wavelength is correspondingly shorter. This ensures that the fastest path is also the one with the fewest wave crests and troughs, as is the case with the shortest path for light reflection at a mirror.

Even more can be shown from these considerations. It doesn't necessarily have to be the shortest (or rather, the fastest) way – the longest (or more generally, the slowest) way would also be okay, if there is such a way. It is only important that the waves at their common endpoint should have about the same wave amplitude for adjacent paths.

So light not only takes the fastest (or slowest) way, but also explores all other ways, and only if we take them all into account will Fermat's principle be derivable as a consequence of interference. If, for example, a cover is glued to the mirror with a very thin slit at the reflection point, then almost all paths – even the closest ones – will be missing. The reflected light then no longer forms a clearly limited beam, which obeys the law of reflection, but is diffracted in different directions, as it would after passing through a narrow slit.

What does this have to do with Feynman and his approach to quantum mechanics? It shows how a classical geometric description by light rays can be derived from a wave model. Does anything similar apply to classical mechanics with its particle paths? Could mechanics also result from a wave description, and what would that mean for this wave model, i.e., for quantum mechanics?

Well, we will soon find out! But before we get there, let's take a closer look at the principle of least action.

1.3 Mechanics Seen in a Different Way: The Principle of Least Action

How does a cannonball, an arrow, or a planet move? Do the same laws of motion apply in heaven as on Earth? These questions were discussed intensively up to the early modern age, and there were several different opinions. Obviously, it is not so easy to develop suitable physical quantities and to find the right laws of motion. In our environment, objects are usually subject to many external influences, such as gravity or frictional forces, making it difficult to get an accurate grasp of the physical situation.

How Objects Move: Newton's Laws of Motion

To illustrate these problems, Feynman proposes the following fictitious law of motion in his Feynman Lectures (Vol. I, Chap. 12.1): *An object left to itself keeps its position and does not move, then when we see something drifting, we could say that must be due to a gorce.* The word *gorce* is of course meant to be reminiscent of *force.* One could imagine, for example, that a locomotive is constantly pulling on some wagons and keeps them moving. Johannes Kepler also had a similar idea around 1600, when he tried to explain the motion of the planets around the Sun. He suggested that the Sun pushes the planets forward on their orbits like a paddle wheel by exerting an *anima motrix (motive soul)* on them, which of course sounds better than gorce.

But ideas like this don't work – they can't be translated into working physics because they contain a fundamental error: they assume that an object will stop moving as soon as there is no external influence on it. Today we know the right law:

> If there is no external influence on a body, it persists in its state of being at rest or of moving uniformly straight forward at constant speed.

You probably recognize it: Feynman's father Melville explained this law of inertia to the young Richard through the example of the ball in the wagon. It was first formulated by Galileo Galilei in 1638. In 1687, almost 50 years later, Isaac Newton adopted this law as the first law of motion in his revolutionary work *Philosophiae Naturalis Principia Mathematica*, in which he established the modern foundations of mechanics. Today, it is generally known as *Newton's first law*.

Now we know what happens without an external influence: the speed of an object doesn't change. Conversely, it implies what should happen when there is an external influence: the speed must change, i.e., the body will accelerate. Note that this does not necessarily mean there will be an increase in speed, because there may be a deceleration or simply a lateral deflection (Fig. 1.14).

With this idea, Newton could finally understand the motion of the planets: they do not have to be driven around the Sun at all, because a constant drive is not necessary according to the law of inertia. Rather, the Sun draws them towards itself via gravity, and thereby guides them along a circular or elliptical path.

In order to calculate the shape of the planetary orbits, Newton had to specify exactly how the Sun causes the acceleration of the planets. For this purpose he used the term *force* and claimed in his second law of motion:

> The alteration of motion (i.e., the acceleration) is proportional to the motive force impressed on the body, and happens along the straight line on which that force is impressed.

Now Newton only had to guess how strong the gravitational force of the Sun must be in order to reproduce the elliptical planetary orbits already found by Kepler. The result was his now famous law of gravity, which states

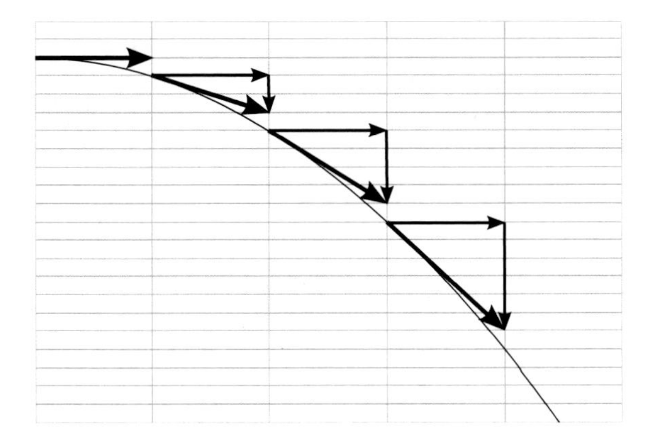

Fig. 1.14 Under the influence of gravity, a freely falling body is accelerated downwards. Its vertical speed component grows steadily, while its horizontal speed component will not change

that the gravitational force is directed towards the Sun and decreases in inverse proportion to the square of the distance from the Sun.

Today, we usually state Newton's second law of motion in short as *force equals mass times acceleration*, formulated mathematically as $F = m \cdot a$, where the acceleration is in the same direction as the force. The mass m comes into play as a constant of proportionality, and indicates how large the inertia of the accelerated body is.

Basically, the law of inertia follows from the second law of motion, because without force the acceleration is zero, so the speed is constant. Nevertheless, Newton listed the law of inertia separately, because only if you recognize the law of inertia can you also find the second law of motion.

Incidentally, this form of Newton's second law is only an approximation that applies at speeds well below the speed of light, and that is of course perfectly sufficient for planets. At higher speeds, force and acceleration are no longer proportional to each other. The acceleration is in fact lower because the object's inertia increases. This is also the reason why nothing can exceed the speed of light. As the speed of an object approaches the speed of light, its inertia becomes greater and any further acceleration under the same force will thus be reduced. Near the speed of light, it becomes close to impossible to accelerate the object any further. Newton could not yet have known all this, because it was only in 1905 – more than 200 years later – that Albert Einstein formulated the correct relativistic law of motion within the framework of his special theory of relativity (see Infobox 1.4).

Infobox 1.4: Albert Einstein's relativistic law of motion

In general, both Newton's non-relativistic and Einstein's relativistic law of motion can be expressed in the following form: the rate of change of momentum p of a body is equal to the force F applied.

The momentum p of the body represents something like "inertia in motion" given by the product of the inertia and the speed. The greater the inertia of a body and the faster it moves, the greater its momentum.

In the non-relativistic case, when the velocity v is well below the speed of light, the inertia is simply the constant mass m of the object, so the momentum is given by the product of mass and speed: $p = m \cdot v$. The rate of change of the momentum is then equal to $m \cdot a$ (mass times acceleration), resulting in Newton's law of motion $F = m \cdot a$.

At higher velocities the inertia of the object increases as its speed approaches the speed of light. Mathematically, this is expressed by multiplying the constant mass m by the speed-dependent Lorentz factor γ, so that the inertia is given by the product $m \cdot \gamma$. The relativistic momentum is thus $p = m \cdot \gamma \cdot v$.

This Lorentz factor γ makes the difference between Einstein's and Newton's laws of motion. It depends on the velocity v of the object in the following way (where c is the speed of light):

$$\gamma = \sqrt{\frac{1}{1-(v/c)^2}}$$

At low speeds γ is almost equal to one and can be omitted. You can easily check this yourself with a calculator. For example, suppose v is the speed of an aircraft, which moves at approximately the speed of sound (about 300 m/s). As fast as this may seem to us, light is still one million times faster, i.e., 300,000 km/s. The value of the Lorentz factor is then 1.000 000 000 000 000 000 000 5.

However, when we approach the speed of light c, the Lorentz factor γ tends to infinity, whereupon the inertia $m \cdot \gamma$ of the object will also grow beyond all limits (Fig. 1.15). If you like, you can calculate it yourself. At 99% of the speed of light (i.e., $v/c=0.99$), you will get a value of approximately 7 for the Lorentz factor, i.e., the inertia will already have increased sevenfold.

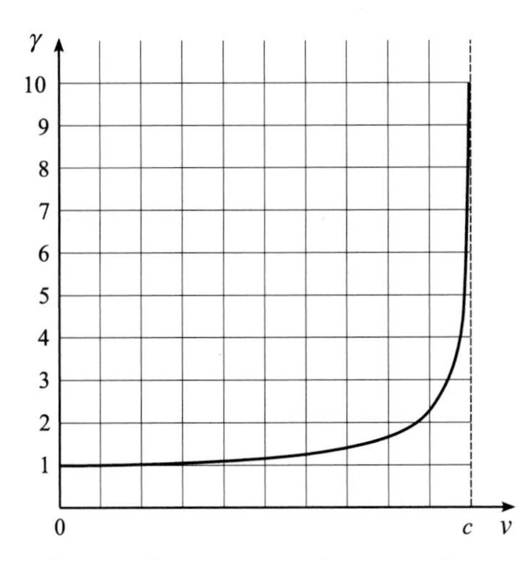

Fig. 1.15 The Lorentz-factor γ becomes ever larger as the speed v approaches the speed of light c. Source: https://commons.wikimedia.org/wiki/File:Lorentz_factor.svg, public domain

The Principle of Least Action – A Cornerstone of Physics

In the 150 years after Newton's discovery, mathematicians such as Leonhard Euler, Joseph Louis Lagrange, and William Rowan Hamilton worked on his equation of motion and found a multitude of different methods for analyzing motion in a universal and elegant way – we have already talked about the Lagrange method, which the young Feynman did not like so much when he first encountered it. In their investigations, mathematicians came across a reformulation of the law of motion which has become known as the *principle of least action*. And as we know, Feynman had been fascinated by the beauty and simplicity of this principle when his physics teacher Abram Bader had shown it to him at high school.

We have already briefly introduced the principle of least action. It is very similar to Fermat's principle, according to which rays of light take the fastest path between two points A and B. This time, however, we are not interested in the path taken by light, but in the motion of a body under the influence of forces such as gravity, in which frictional forces are omitted.

In order to apply the principle, we must also consider, in addition to the actual motion, all possible imagined motions between the two points, including those which do not obey Newton's law. There is only one condition: the time it takes to get from A to B must always be the same. The body should therefore start at position A at a certain time t_1 and arrive at position B at another fixed time t_2 (Fig. 1.16). What distinguishes the actual motion from all the imagined motions?

Analogous to the Fermat principle, there is a certain quantity which can be calculated for each of these motions and which becomes minimal for the actual motion. This is the so-called *action S*. For this purpose, the available time is divided into tiny time intervals dt, so that the kinetic energy $T = m \cdot v^2/2$ and the potential energy V are practically constant during these short lapses of time, although they can change from one such time interval to the next. We calculate the difference $T - V$ of the two forms of energy for each time interval, multiply this difference by the duration dt of the interval, and add the results for all the time intervals to obtain a total sum.

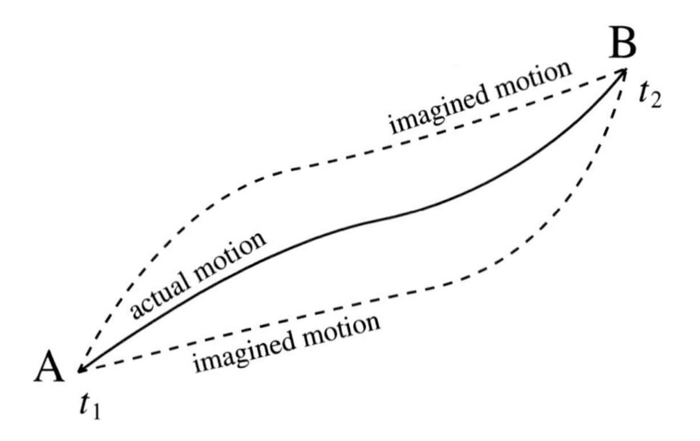

Fig. 1.16 Actual and imagined motions of a body between points A and B

For each motion from A to B, we can calculate this action. It says to a certain extent how “effective” or “favorable” the corresponding motion is. The greater the potential energy V of the body and the lower its kinetic energy T on average, the more favorable the motion will be. In any real situation, the body will “choose” the most favorable motion, i.e., the one with the least action. According to the principle of least action, the object tries to move between the two points A and B in such a way that on average its kinetic energy T becomes as small as possible and the potential energy V as large as possible.

So it would be optimal for the kinetic energy if the motion of the body were as slow as possible. However, the object must arrive on time at location B, at the fixed time t_2 – this was our requirement for all the motions considered. Therefore the object cannot move arbitrarily slowly. Furthermore, it should try to get into an area where the potential energy is as high as possible and stay there as long as possible.

Let us look at a simple example. We throw a ball vertically into the air at position A and demand that it arrives a certain time later at position B, where B should lie vertically above A. In order to increase its potential energy V, the ball will try to move as far up as possible, where the potential energy is greater than below. (The gravitational potential grows in proportion to the height x according to the formula $V = m \cdot g \cdot x$, where g is the gravitational acceleration.) The ball tries to stay there as long as possible and then fall down to point B. However, it can't move too far up, because then it would have to go quite fast to reach B in time, and this would increase its kinetic energy. The exact details depend on how much time is available for the flight from A to B. If there is plenty of time, the ball will move far up and use the high potential energy there. If it doesn't have much time, however, it will stay further down in order not to move too fast.

Obviously, we can understand how the ball will move using the principle of least action. The exact calculation shows that the motion with the least action will be a parabola in a path versus time diagram (Fig. 1.17). Newton's law of motion would give the same result. This must of course be the case, because Newton's law and the principle of least action are equivalent ways to describe the motion of objects.

It is even possible to derive Newton's law of motion mathematically from the principle of least action. The calculation is not difficult, but it goes beyond the scope of this book – you can find it, for example, in the *Feynman Lectures on Physics*, Vol. II, Chap. 19.[6] If you use general

[6]The Feynman Lectures are freely available on the Internet at http://www.feynmanlectures.caltech.edu/.

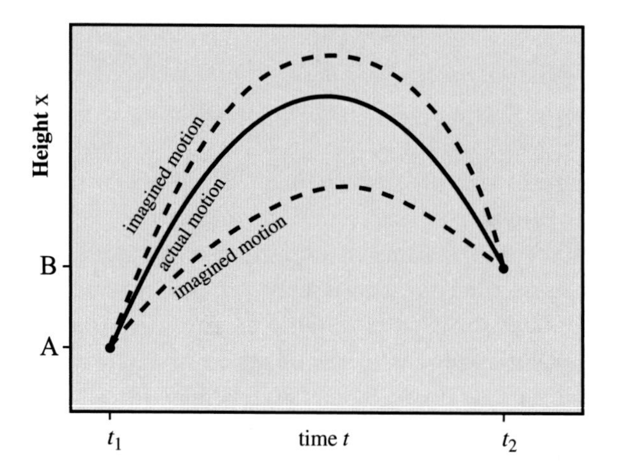

Fig. 1.17 Vertical rise and fall of a body between points A and B (time increases to the right)

coordinates that are optimally adapted to the corresponding motion, you can even derive the elegant Lagrangian method from Infobox 1.2.

In his lecture entitled *The Principle of Least Action*, Feynman also presents an argument as to why something like Newton's law of motion should arise from an apparently completely different formulation, such as the principle of least action. This is typical of Feynman. He was never content with a formal derivation, but always tried to find out what was actually going on behind it. Why does the calculation work?

Here is his argument. If one considers the actual motion between points A and B in the given time, the principle of least action does not only apply to the complete motion. We can pick out any part of the motion from a point A′ to another point B′, both of which lie on the path of the object – this part will also be a motion with least action, because it has the smallest action of all motions between A′ and B′ with the same running time. That must be the case, because this part is also an actual and not an imagined motion.

We can now make the selected part of the motion smaller and smaller, until it becomes infinitesimally small. In this way, the spatial derivative of the potential energy and the time derivative of the kinetic energy now come into play, and this ultimately leads to force and acceleration. We thus recover exactly the terms appearing in Newton's law of motion.

Feynman's argument makes it clear why it is possible to go from an integral description (least action), which covers the entire motion, to a differential description (Newton's law of motion), which only requires the values and rates of change of physical quantities at one point in time. This is due to the fact that the integral description also applies to arbitrarily small parts of the motion. In this way, we can also switch between a big picture involving integrals and a detailed description in terms of differentials in many other areas of physics.

The principle of least action is far more general than it seems here. It is not limited to mechanics, but can also be applied to the electromagnetic field, for example. The action is then no longer simply the time integral of $T - V$ for the motion, but a space-time integral of a more complex function of the energies and fields, which by analogy with the Lagrangian function $L = T - V$ is called the *Lagrangian density*. In fact, every fundamental law of nature known today can be expressed by a suitable action, whose minimization results in the corresponding equations of motion or field equations.

I would like to make one final comment. In Fermat's principle we have seen that the travel time of a light beam does not necessarily have to be minimal. It can also be maximal, or more generally, stationary, so that the light travel times of neighbouring paths differ only very little. For the principle of least action we have the same: a maximal action would also work. Nevertheless, we shall continue to talk simply about the least action – and in the vast majority of cases it will be true!

It is clear that there are many similarities between Fermat's principle for light rays and the principle of least action in mechanics. We might therefore expect the two principles to have a common basis, namely the wave nature of radiation and matter. It was Richard Feynman who succeeded in developing an alternative view of quantum mechanics on the basis of this idea. Find out more in the next chapter.

2

Princeton, Path Integrals, and the Manhattan Project

The move to Princeton was the beginning of an intense period for Feynman. He succeeded in finding a completely new way of looking at quantum mechanics. He also took part in the Manhattan project, the USA's attempt to beat Germany to the construction of a nuclear bomb during the Second World War. And finally he had to cope with a private tragedy, because Arline, his childhood sweetheart and later wife, died of tuberculosis shortly before the end of the war.

But before we turn to these dramatic events, let's take a look at how the unconventional Feynman found life and work at the very traditional Princeton University.

2.1 Feynman in Princeton

In the fall of 1939, Feynman's father Melville drove his son to the prestigious Princeton University, which is located about two hours' drive southwest of New York. With its neo-Gothic buildings, Princeton looked like an imitation of the elite English universities of Oxford and Cambridge. One could even hear the British accent when tea was served in the afternoon. There was a porter downstairs, and during meals together in the great hall they wore academic gowns. If you think of Hogwarts School of Witchcraft and Wizardry from the Harry Potter films, you won't be so far from the truth (Fig. 2.1).

J. Resag, *Feynman and His Physics*, Springer Biographies,
https://doi.org/10.1007/978-3-319-96836-0_2

Fig. 2.1 Alexander Hall at Princeton University. Source: https://commons.wikimedia.org/wiki/File:Princeton_University_Alexander.jpg

So Princeton wasn't exactly the ideal place for an unconventional person like Feynman. But Feynman decided to be "nice", as he writes in *Surely You're Joking, Mr. Feynman*. So he did his best to adapt as far as possible to the local rules and customs, something that was not always easy for him – after all, he hated all kinds of formalities. Over time, however, he realized that some of the conventions had their advantages. For example, you didn't have to change your clothes for dinner – you just grabbed your academic gown, and put it over the top of whatever you were wearing.

What particularly impressed Feynman was that Princeton had a cyclotron – a small particle accelerator by today's standards – which produced many important experimental results. When Feynman found it in a basement room of the physics building, he knew he was in the right place at Princeton: that was physics as he liked it. The room was crammed full of all kinds of equipment, cables, and switches, and Feynman immediately felt as though he was back in his own little home-built laboratory from childhood days. In the midst of all this chaos, it was only possible to keep track of things if you had been working actively with the devices for some time and knew exactly what you were doing. Anyway, by this time Feynman was happy that Slater had told him to leave MIT and see what the rest of the world had to offer.

Princeton had another advantage: apart from physics, which we are going to talk about in a moment, Feynman had many opportunities to come into contact with other disciplines. There were no compulsory subjects as at MIT – for which Feynman was very grateful – but every Wednesday a wide variety of people came to Graduate College and gave lectures on topics such as poetry, religion, and hypnosis. Of course, Feynman was one of the first to be hypnotized, and indeed – it worked! In Surely You're Joking, he describes his experience like this: You're only slightly fogged out, and when the hypnotist says that you can't open your eyes, you're pretty sure you could open your eyes. But of course, you're not opening your eyes, so in a sense you can't do it.

Feynman and the Philosophers: When Is an Object Real?

Since Feynman was interested in a variety of subjects, he made it a rule when sitting in the large dining hall not only to consort with physicists, but also to get to know other groups. He did not shy away from the philosophers either, and so came to have an experience that reconfirmed his prejudices against philosophy. One day he got into a discussion about the book *Process and Reality* by the British philosopher and mathematician *Alfred North Whitehead*. Feynman was invited to a seminar about this book and promised himself to keep his mouth shut – after all, he wasn't exactly an expert in philosophy. In the seminar, the term *essential object* was often used, and after some discussion, the professor turned to Feynman and asked him: "Would you say an electron is an essential object?" Feynman had no idea what an essential object might be, and so he replied with a counter-question: "Is a brick an essential object?" A wild discussion arose in which the most diverse views were expressed: Was the individual brick an essential object, or did this term refer to the general character that all bricks have in common (their 'brickiness'), or was it the idea in the mind that you get when you think of bricks, and so on. The discussion ended up in complete chaos. From Feynman's point of view, this was typical of philosophical discussions – they didn't even have a clear answer to such a fundamental question, so what sense could be made of a seminar on the whole book?

Feynman himself would have answered the question of the brick or electron by saying that both are theoretical constructs that we use to better understand nature. They are so useful that we can call a brick or an electron *real* in this sense. The theoretical construct of a brick, for example, includes the fact that it has an inside, but this cannot be said of the electron if we go by today's understanding. However, nobody has ever seen the inside of a brick, because if you break it, you only see a surface of the brick again. The situation with the electron is similar when one speaks of properties such as its mass, energy, momentum, location, spin, or charge. All these are conceptual constructs, which make sense within the framework of certain physical theories and thus help us to get a better understanding of matter. But no one has ever been able to touch or see a single electron in the same way as we can with a brick.

The views on what can be considered *real* in physics have changed over time. Around 1900, for example, the Austrian physicist and philosopher Ernst Mach called the real existence of atoms into question, asking flippantly: "Ham se welche gesehn (have you seen any)?" Today, atoms have become indispensable for describing nature, and every scientist would accept them as real objects without hesitation.

The question of whether a physical term represents something real will be encountered in several places in this book. For example, is an electromagnetic field real? Many physicists would shout YES, since such a field can be assigned an energy and momentum density, for example. Nevertheless, electrodynamics can be completely formulated without the field concept, as we will soon see. But this would have been no problem for Feynman, because a field was only a theoretical construct for him, and it was its usefulness for understanding nature that determined how real it could be considered to be.

Wheeler Has Always Been Crazy

So Feynman didn't really warm up to philosophy in Princeton. This was completely different with physics! Princeton had offered him a position as an assistant to Eugene Wigner, and this provided him with some financial security – much to the relief of his father, who was suffering increasingly from health problems and feared that at some point he would no longer be able to support his family. Although Wigner was undoubtedly an outstanding physicist, it was fortunate for Feynman that he was in fact assigned to John Archibald Wheeler instead (Fig. 2.2). Wheeler was only seven years older than Feynman, and proved to be a perfect associate for Feynman, because although he presented himself as a perfect gentleman, in perfect keeping with tradition, he was actually one of the most creative minds in modern physics. He did not shy away from bizarre ideas, whether they were parallel universes, quantum foam, wormholes, or black holes – and by the way, the last three terms were coined by him. Feynman later said about

Fig. 2.2 The young John Archibald Wheeler (1911–2008) (© Emilio Segre Visual Archives/American Institute of Physics/Science Photo Library)

Wheeler: "Some people think Wheeler's gotten crazy in his later years, but he's always been crazy!"[1]

At the outset, however, it didn't look as if Feynman and Wheeler would get along very well with each other. Wheeler was still young and maybe wanted to make sure that Feynman would respect him. So in their first meeting he laid out a pocket watch on the desk so that they could stick to the allotted time. One can imagine what this formal and rather condescending manner must have done to the unconventional Feynman. He bought a cheap pocket watch, and when Wheeler placed his watch on the desk again at the next meeting, Feynman did the same, ostentatiously demonstrating that his time was just as valuable as Wheeler's. This could have gone wrong with some other professors, but Wheeler was by no means the pompous busybody he appeared to be. Both recognized the amusing aspect of the situation and burst out laughing. The ice was broken and there developed a long-lasting friendship and fruitful collaboration between these two brilliant physicists.

[1]Quoted in Dennis Overbye: *John A. Wheeler, Physicist Who Coined the Term Black Hole, Is Dead at 96*, New York Times (April 14, 2008).

2.2 Electrodynamics Without Fields

The first tasks Wheeler assigned to his new Ph.D. student Feynman were closely related to Wheeler's own research. Wheeler had dealt intensively with the quantum mechanical description of scattering processes, in which different particles collide with each other. Feynman learned a lot about quantum mechanics from this. But he also had his own ideas and began to think again about one that had already occupied him at MIT.

Mysterious QED

The background of Feynman's idea was this. At that time, physicists were making great efforts to combine quantum theory with electrodynamics in order to derive a new theory called quantum electrodynamics (QED). But the Schrödinger equation would not work for this purpose, since it was limited to the non-relativistic case where speeds were far below the speed of light. This was usually accurate enough to describe electrons in the shells of atoms. However, electrodynamics also incorporates light waves, and these must of course move at the speed of light. So a description was required that was consistent with Einstein's Special Theory of Relativity.

We have already discussed two relativistic quantum equations: the Klein – Gordon equation and the Dirac equation. But these equations were not sufficient to achieve the goal. Although they could describe the relativistic quantum motion of particles such as electrons, the electromagnetic forces acting on them were still represented by a classical electromagnetic field.

Why should that be unsatisfactory? The crucial point is that electromagnetic fields do not have to be static. For example, if an electric charge in an antenna oscillates back and forth, the fields vary periodically and can radiate out from the antenna as an electromagnetic wave. Light is just such an electromagnetic wave. Thanks to Planck and Einstein, it was known that the description of light by such a wave was only a classical approximation. When we look closely, light consists of a stream of particles, called photons.

It was clear that electromagnetic waves would also require a quantum description that linked the wave aspects of light with its particle character, as happened with electrons and all other particles. Since electromagnetic waves were only a special case of electromagnetic fields, it seemed logical that all electromagnetic fields should be described quantum mechanically by photons. For example, the electrical attraction of the atomic nucleus on the electrons in an atom could then no longer be represented by a classical electric

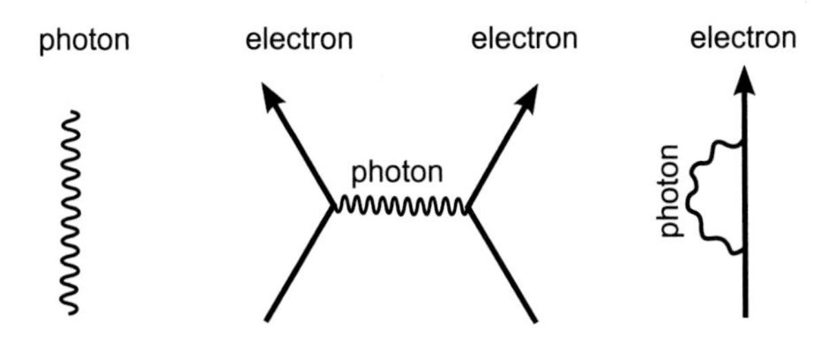

Fig. 2.3 Quantum theory describes electromagnetic interactions in terms of photons. These can move as free particles (left), act between charged particles (center), or cause self-interaction of a particle (right)

field, but had to be described by the effect of photons that somehow moved back and forth between the atomic nucleus and the electrons.

The formulation of quantum electrodynamics using photons proved to be extremely complex, and serious difficulties were encountered, which fundamentally called the whole project into question. As long as only simple calculations were made, in which only one photon was exchanged between, e.g., the atomic nucleus and an electron, reasonable results could be obtained. However, as soon as several photons came into play at the same time, the results were often useless. For example, the charge of the electron seemed to become infinitely large.

Another problem was the self-interaction of the electron. If an electron emits a photon, the photon does not necessarily have to move towards the atomic nucleus. It can also follow a loop back to the electron and be captured by it again. The electron thus interacts with itself by emitting and absorbing the same photon (Fig. 2.3).

Now the photons involved in a self-interaction have a certain energy, which is located near the electron, and according to Einstein's Special Theory of Relativity, a localized energy corresponds to a mass. Indeed, if we add the energy E to an object at rest, its mass will increase by E/c^2. Conversely, its mass will shrink by this amount when it releases the energy E. There are even processes in nature where the whole mass m of a particle is released as energy according to Einstein's formula $E = m \cdot c^2$. One example is the radioactive decay of a neutral pion into two photons, which still have energy and momentum, but no longer have any mass (see Infobox 2.1).

In QED, a real electron always consists of the bare electron plus a cloud of photons, which it constantly emits and absorbs. The energy of this photon cloud must be included when talking about the mass of a real electron. And this is where the problem arises: the calculations yield an infinitely large amount of energy for the photon cloud, whence the mass of the real electron would have to be infinitely large. But that can't be true, of course! Furthermore, the problem is not caused by quantum theory. Even in classical electrodynamics, the energy of the electric field of a point-like particle is infinitely large, as we shall see.

Infobox 2.1: What is the momentum of a particle without mass?

How do photons manage not to have a mass, and yet still carry energy and momentum, enough to be able to push electrons out of a metal surface, for example?

In classical Newtonian mechanics such particles could not exist, because a particle without mass would have no energy and no momentum. It would thus have no inertia at all. In that context, such a particle would make no physical sense.

This changes when we consider the theory of special relativity. In Sect. 1.3, we saw that the inertia of a particle is not only due to its constant mass. Rather, it becomes greater as its speed approaches the speed of light.

Now consider the following question: what must be done to ensure that a particle retains a certain inertia, even though we make its mass smaller and smaller? The answer is that it must move faster and faster, coming ever closer to the speed of light, so that the increase in speed exactly compensates the decreasing mass and the inertia remains constant. In the limit of infinitely small mass, the speed will be infinitely close to the speed of light, so we may say that massless particles must always fly at the speed of light. This is not only true for photons, but for all massless particles.

So the idea of a massless particle makes sense thanks to relativity theory! Such particles move at the speed of light and have inertia, so they also have energy and momentum.

Feynman had learned about the difficulties involved in the formulation of quantum electrodynamics when he studied the relevant books by Heitler and Dirac at MIT. At that time he was not yet able to understand the technical details of the calculations, but he nevertheless got some insight into these problems. In particular, the following sentence from Dirac's book *The Principles of Quantum Mechanics* stuck in his mind: "It seems that some essentially new physical ideas are here needed."

Electrons Without Self-interaction – Is that Possible?

At that time, an idea was maturing in Feynman's mind which, from today's perspective, led only to a minor branch of scientific development.

Nevertheless, since it finally motivated Feynman to formulate a completely new approach to quantum theory, we shall take a closer look at it. The basic idea is as follows. The cause of the problems with QED seemed to be that there could be any number of photons involved in a process and that the electron could interact with itself, i.e., with the photon cloud it emits and absorbs all the time. But why should a particle interact with itself at all? This seemed absurd to Feynman, and indeed unnecessary, because charged particles should only interact with other charged particles, not with themselves. Wasn't it possible to reformulate the classical theory in such a way that the self-interaction of electrons would disappear and they would only interact with other electrons? But how could that be achieved?

Well, we would have to do without the usual electric (and magnetic) field. Normally, the procedure is to calculate a combined field due to all the charges present, where each charge contributes to the field. At the same time, each charge is affected by the field due to all the charges, including its own contribution. So if we use a single combined field, the self-interaction is unavoidable.

But can we really do without such electric and magnetic fields? Aren't they an element of reality? Doesn't light consist of an electromagnetic wave? Even Albert Einstein once said: "For the modern physicist, the electromagnetic field is no less real than the chair on which he sits."

But we have to be careful here! Such considerations are very much reminiscent of the discussion about *essential objects* that Feynman had witnessed among the philosophers at Princeton. According to Feynman, electromagnetic waves were merely theoretical constructs that were useful for describing reality – just like Einstein's chair, by the way. Instead of saying that we detect light with our eyes, we can also say that we perceive the electromagnetic influence of a luminous object on the sensory cells of our eye. Light is always emitted from some object, and this object is exactly what we see. To begin with, for example, electric charges oscillate in the Sun, and eight minutes later other charges in our eyes are caused to vibrate in consequence, something which we perceive as brightness. Perhaps we can therefore dispense with light and other electromagnetic fields as mediators and describe the influence of charges on each other directly, this taking place with a certain time delay corresponding to the light propagation time.

When physicists first formulated mathematical laws for forces between static charges, they proceeded in a similar way. This concept is called *action at a distance*. Newton's law of gravitation uses exactly this concept, according to which two masses attract each other but we do not specify how this force

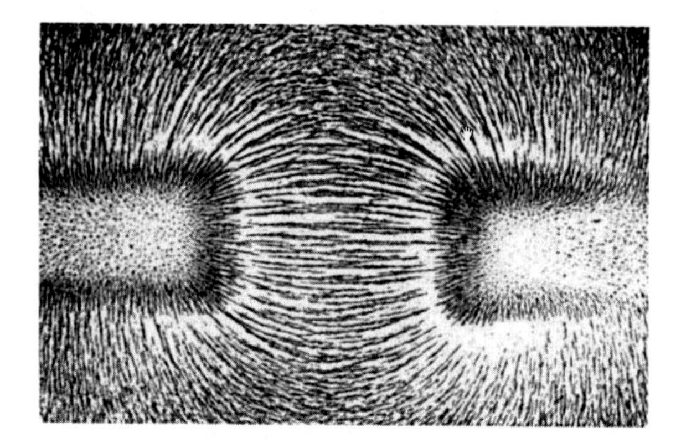

Fig. 2.4 Iron filings orient themselves parallel to the magnetic field lines at each point. Source: https://commons.wikimedia.org/wiki/File:Magnetic_field_of_bar_magnets_attracting.png. *Credit* Alexander Wilmer Duff

is transmitted between the two masses. Even Newton himself had expressed concerns about this and hoped that at some point someone would explain how the force was actually transmitted.

Around 1852, the English physicist Michael Faraday had introduced the terms "electric" and "magnetic field". He had imagined rather vividly that lines of force would penetrate space as though it were a medium and have an effect on the movement of charges (see Fig. 2.4). For him, space was filled with invisible field lines which created the forces on charges. As James Clerk Maxwell described it in 1873[2]: "Faraday, in his mind's eye, saw lines of force traversing all space where the mathematicians saw centres of force attracting at a distance; Faraday saw a medium where they saw nothing but distance."

It was in fact James Clerk Maxwell who, in 1864, on the basis of Faraday's work, succeeded in summing up the complete interrelationship between fields and charges in a few mathematical equations (see Infobox 2.2). In these equations the electromagnetic field is so useful as an intermediary for the forces between charges that we often consider it indispensable, and therefore as real. But the whole system could also work without this mediator!

[2]James Clerk Maxwell: *A Treatise on Electricity and Magnetism*. Clarendon Press, 1873.

Infobox 2.2: The Maxwell equations

The Maxwell equations describe the mutual interrelationships between electromagnetic fields and electric charges and currents. Electrically charged particles such as electrons, for example, generate an electric field in their environment, i.e., the charges act as sources for the electric field. In contrast, magnetic fields do not have such sources. There are no magnetic charges – they would be called magnetic monopoles if they existed. Therefore every magnet has a north and a south pole and there are no north or south poles moving around alone.

Generally, magnetic fields are generated by electric currents. Each current-carrying cable generates such a magnetic field, which can be detected with a compass needle, for example. If you wind the cable into a coil, you can make an electromagnet with it.

On the other hand, a magnetic field that changes over time causes an electric current in a cable. Strictly speaking, the variable magnetic field first generates an electric field, which then sets the electrons in the cable in motion. In this way, for example, we can build a dynamo to generate electricity.

Maxwell's particular insight was that variable electric fields also generate magnetic fields. This creates a fascinating possibility: an oscillating electric field can generate an oscillating magnetic field, which in turn generates an oscillating electric field and so on. The oscillating electric and magnetic fields can thus detach themselves from the charges and propagate as electromagnetic waves into space, where they keep each other alive, as it were. Maxwell used his equations to predict the existence of electromagnetic waves, and this was most impressively confirmed in 1886 by the German physicist Heinrich Hertz in a famous experiment.

Feynman hoped that, by avoiding a general electromagnetic field, all problems could be solved, even in the corresponding quantum theory: the electron would no longer interact with itself, so there would no longer be an infinite self-energy.

At least, that was the plan, as Feynman wrote in his Nobel Prize speech: first solve the classical problem, get rid of the infinite self-energies in the classical theory, and hope that when you made a quantum theory of it, everything would just be fine. That was the beginning, and the idea seemed so obvious to Feynman and so elegant that he fell deeply in love with it.

The Radiation Resistance of Oscillating Electrons

In Princeton, he set about developing this idea. He had learned a lot during the previous years and had come up against a problem that threatened to call his wonderful idea into question. When an electron oscillates back and forth in an antenna, it emits electromagnetic waves and thus energy. So the electron is constantly losing energy while oscillating. This dampens the motion, rather as it would due to a frictional force. This effect is called *radiation resistance*. In order to maintain the oscillation, the radiation resistance must be counteracted, somehow supplying the radiated energy back to the electron.

However, the comparison of the radiation resistance with a frictional force is not quite correct, because there is an important difference: a frictional force is related to the speed of an object, whereas radiation resistance is caused by the changing acceleration during the oscillation. A uniformly moving electron does not radiate, and nor does a uniformly accelerated electron. The radiation resistance can only be felt when the acceleration changes, as with an oscillating electron.

The Dutch physicist Hendrik Antoon Lorentz had investigated the origin of this radiation resistance, and had come to the conclusion that it lay in the self-interaction of the electron. However, this was just what Feynman was trying to get rid of – his basic idea was to say there is no such thing as the electron acting upon itself, because this leads to an infinite self-energy and thus to an infinite mass for the electron. So Feynman was just about to throw away the baby with the bath water.

It is very instructive to take a closer look at Lorentz's train of thought regarding the self-energy and radiation resistance of the electron. The basic idea is as follows.[3] Lorentz did not visualize the electron as a point, but assumed a small spherical charge distribution. A simple model, for example, is a sphere on whose surface the negative charge of the electron is distributed evenly. Each individual area of the negatively charged sphere surface repels all the others – this is the self-interaction of the sphere. Because of this repulsion, energy would have been needed to transfer the charge from some distant point to the surface of the sphere, and this energy is stored in the electric field, where it contributes to the mass of the real electron according to Einstein's formula $E = m \cdot c^2$.

If we now shrink the sphere, we have to work against the electrical repulsion of the charge on its surface – so we need energy to reduce the size of the sphere. The smaller the sphere, the stronger the repulsion becomes, so that more and more energy is stored in the field. Correspondingly, the mass equivalent of this energy continues to increase. If the sphere shrinks to a single point, the energy and thus the mass of the sphere will become infinitely large. So a point-like electron should have an infinite mass and energy, caused by the interaction of its charge with its own field. This was precisely why Feynman wanted to get rid of the self-interaction.

---

[3]See Feynman Lectures, Volume II, Chap. 28.

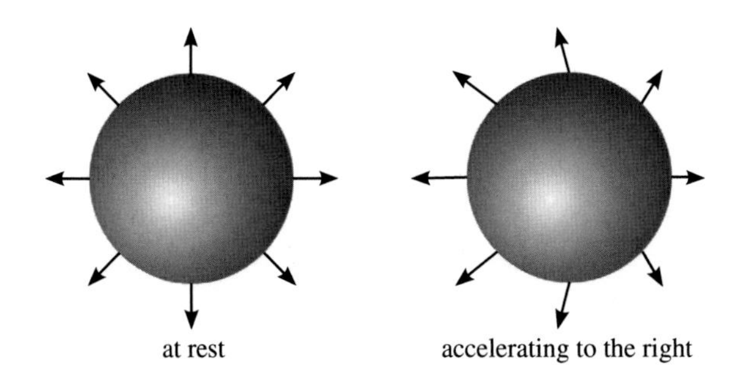

Fig. 2.5 The self-force on an accelerating electron is not zero

But maybe the electron isn't point-like at all. So how large would the sphere have to be for the whole mass of the electron to be caused by its electric field alone? The calculation is straightforward, and it turns out that the sphere would have to be about as big as a proton, i.e., roughly $1\ \text{fm} = 10^{-15}$ m. However, there is still no experimental evidence that the electron has any spatial extension. The upper limit is currently about 10^{-19} m, i.e., the electron has to be at least ten thousand times smaller than the model of the sphere allows.

Thus, the model cannot correspond to reality in this form. Classical electrodynamics is not compatible with the sphere model of charged particles that are very small or even point like. Nevertheless, the sphere model is quite capable of explaining the radiation resistance mentioned above, as we will see now.

As long as the sphere is at rest or moves at constant velocity, the forces of mutual repulsion on different parts of the surface of the sphere balance each other exactly. But this changes when the sphere is accelerated, because the influence of one part of the sphere takes time to affect the other parts. According to relativity theory, electrical forces can only propagate at the speed of light. The ever-changing positions of the individual parts of the sphere only have a delayed effect on the other parts, in such a way that the electrical forces as a whole no longer balance each other out. In short, a force is created that tries to oppose the acceleration (Fig. 2.5). It's as though the electron holds itself back by its own bootstraps.

This self-force can be calculated and broken down into several parts.[4] One part is proportional to the acceleration and increases as the radius of

[4]The exact formula can be found in the Feynman Lectures Volume II Chap. 28-4 Eq. 28.9.

the sphere shrinks. We already know this part: it contributes to the inertia of the sphere, an inertia created solely by the self-interaction of the sphere, and which corresponds to the self-energy stored in the field. The smaller the sphere, the greater the self-energy and hence also the inertia of the sphere, until finally it becomes infinitely large for a sphere of radius zero. This is exactly what Feynman wanted to get rid of by denying the self-interaction.

In addition, there are other components of the self-force. All of these disappear when the sphere shrinks to a point, except one. This remaining part is proportional to the rate of change of the acceleration and corresponds exactly to the radiation resistance needed to explain the energy emission of an oscillating electron. Moreover, it does not depend on the size of the sphere, so the sphere can shrink to a point and there is no problem for this explanation of the radiation resistance. Therefore the sphere model can easily explain the radiation resistance of the electron, even if we assume it to be point-like.

Feynman was therefore faced with a dilemma regarding the sphere model of the electron. On the one hand, the self-interaction leads to an inertia which would be too big if we take into account the maximum size the electron can have according to experiment. This inclines us to get rid of the self-interaction. On the other hand, the self-interaction leads to a radiation resistance which occurs whenever the acceleration changes over time, and this is a real physical effect that would have no explanation without self-interaction.

In the context of classical electrodynamics, there seem to be few alternatives to the sphere model or similar models for the electron. We must somehow visualize the electron and its charge distribution, even if we let the sphere shrink to a point. But then contradictions seem unavoidable: with self-interaction, the inertia of the sphere is too big, and without self-interaction, the radiation resistance is missing. So did that spell the end of Feynman's idea?

If electrons can only act on other electrons, but not on themselves, there is only one solution: the influence of other electrons in the universe on the oscillating electron must be responsible for its radiation resistance. Feynman had the following idea. If the electron vibrates for a short time, it can also cause another electron in its environment to vibrate through their electromagnetic interaction. This other electron will in turn act back on the first electron, and this might explain its radiation resistance. Feynman carried out the calculations, but it didn't look right. So Feynman went to Wheeler and asked him for advice.

The answer Feynman had obtained depended on the mass and charge of the second electron and would also become weaker the further away the two electrons were from each other, whereas the radiation resistance did not have any of these dependencies. In addition, the time aspect of things did not fit. The radiation resistance is immediate, as soon as the first electron begins to vibrate, but it takes time for the vibration of the first electron to act on the second electron and back again, because the mutual influences of the electrons propagate at the speed of light. What Feynman had calculated was just ordinary everyday reflection of light.

So Feynman's idea didn't seem to work. But Wheeler was known for being fascinated by seemingly strange ideas. Therefore he didn't simply reject Feynman's idea, but rather recognized the potential in it and began to wonder what could be done to save the idea.

The dependence on the distance between the two electrons could be removed by assuming that there were enough electrons in the universe to completely absorb the radiated energy of the first electron. If one thought more closely about this, the dependencies on the masses and charges of the other electrons could also be eliminated.

Retarded and Advanced Fields: Forward and Backward in Time

The problem with the time delay remains. Here Wheeler had a seemingly crazy idea: what if the effect of the other electrons back on the first electron did not only propagate forward in time – these fields are called *retarded fields* – but also backwards in time? These would then be called *advanced fields*. The retarded fields or waves represent the usual reflection and scattering of light. The advanced fields, on the other hand, would arrive at the first electron at the right moment, exactly when it starts oscillating, and could thus generate the radiation resistance.

Does that make any sense? How can fields or electromagnetic waves propagate backwards in time? Well, mathematically, the Maxwell equations of the electromagnetic fields are time-symmetrical. They make no distinction between the future and the past. When we record a purely electromagnetic process on a film and then reverse this film, we see another physically possible electromagnetic process.

Nevertheless, physicists normally use only the retarded fields of charges and simply ignore the advanced fields. This is due to the fact that electric charges are normally moved by certain other influences. The charges are then the cause and the reaction of the fields is the delayed effect. One example is provided by the Sun: nuclear fusion generates large amounts of energy in it, and this causes electric charges in its atoms to vibrate, whereupon the atoms emit electromagnetic waves into cold space.

If the Sun also radiated advanced waves into the past, this would correspond in the usual time direction to a process whereby electromagnetic waves from outer space would home in towards the Sun and be absorbed there. The waves would then be the cause and their absorption the effect. The Maxwell equations do not exclude this, but such a thing does not usually occur in our universe because space is cold and dark.

But nobody forbids us to think about advanced fields and waves. In his Nobel Lecture, Feynman describes it like this: "Sometimes an idea which looks completely paradoxical at first, if analyzed to completion in all detail and in experimental situations, may, in fact, not be paradoxical."

What does it look like when a briefly vibrating charge emits both a retarded wave into the future and an advanced wave into the past? In a film running forward in time, the sequence would be as follows (see Fig. 2.6). To start with (large negative times), the advanced wave, in spherical form, closes in the charge from all sides, while the radius of the sphere becomes smaller and smaller. At zero time, the radius of the advanced wave sphere has shrunk to zero and the charge absorbs the advanced wave, causing it to vibrate and emit a retarded spherical wave into space.

This process is completely time-symmetrical. If we let the film run backwards, the sequence looks exactly the same, except that the retarded and advanced waves would change their roles.

Feynman and Wheeler now imagined that the first electron would vibrate briefly in this way at time zero, sending out an advanced and retarded wave. They imagined that it would be surrounded in every direction by many other electrons which would absorb the emitted retarded wave completely at a later time. Each absorbing electron would thus be set into a short vibration

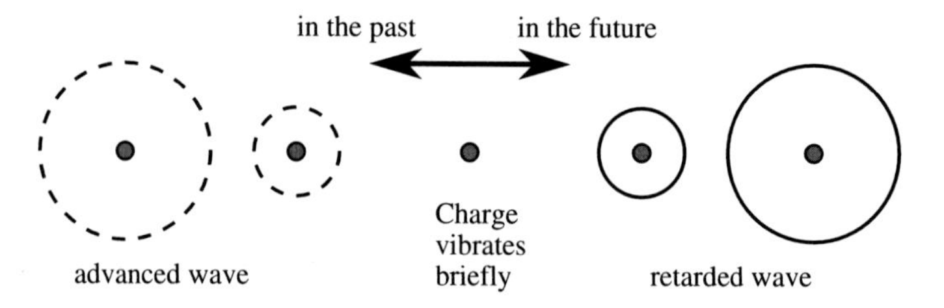

Fig. 2.6 Retarded and advanced wave of a briefly vibrating charge. Time increases from left to right

and also emit an advanced and a retarded spherical wave. The new retarded waves simply correspond to the reflection of the original retarded wave from the first electron.

But what would happen to the newly emitted advanced waves? They would run backwards in time and hit the first electron exactly at the point in time when it emitted its waves. In a forward running film the process would look like this. For times earlier than zero, each electron is surrounded by a shrinking advanced spherical wave (see Fig. 2.7 left). As we approach zero time, the advanced spherical wave of the first electron shrinks to a point. At the same time, the advanced waves of the other electrons approach this electron from all sides. Since there are many advanced waves, they basically add up to a spherical wave that moves synchronously with the advanced and later the retarded wave of the first electron and interfere with it.

And now comes the surprise! Feynman calculated that, for negative times, the advanced waves of the other electrons and the advanced wave of the first electron extinguish each other. For positive times, however, they enhance the retarded wave of the first electron (Fig. 2.7 right). Taken together, it looks as if the first electron did not send an advanced wave into the past at all, but only a retarded wave into the future – exactly as we would expect from a vibrating charge that radiates energy. The advanced response of the other electrons thus ensures that the vibrating electron loses energy and experiences a corresponding radiation resistance – without any self-interaction. The whole approach produces a reasonable overall result!

However, one thing does seem strange. Although the concept is time-symmetrical, the final result seems to prefer a time direction. If all retarded and advanced waves are taken together, the result is that the

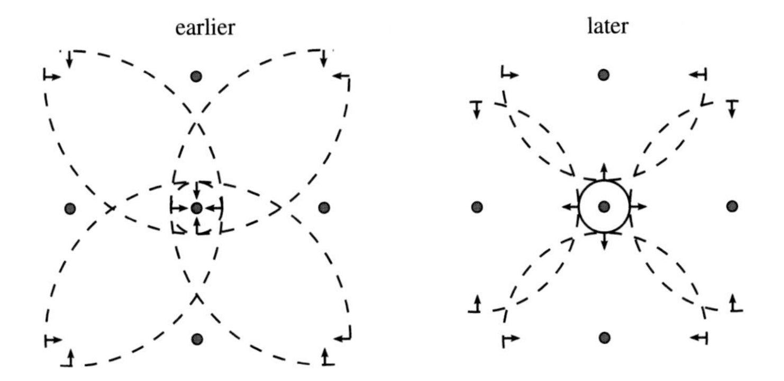

Fig. 2.7 Retarded and advanced waves with four absorbing electrons. For details see main text

vibrating first electron emits a wave into the future that is absorbed by the other electrons. Would it be possible to reverse the process in time, just as our time-symmetrical approach requires? Let's take a look.

If we examine the process backwards in time, we find the following. All electrons except for the first emit a spherical wave, each at exactly the right moment, and these add up to a single spherical wave. This wave contracts around the first electron and is absorbed by it at time zero. This is physically possible, but very unlikely, because it requires the coordinated action of many electrons. This makes it clear where the time asymmetry that we normally observe comes from. The reverse process is not forbidden, but due to the large number of particles involved, it is so unlikely that we will never see it in reality. Thus, the theory of Feynman and Wheeler is exactly in line with other explanations for the formation of a time direction.

So Feynman's and Wheeler's theory looked promising. Feynman put in even more effort to test his ideas with a multitude of examples. What exactly were the requirements for the distribution of the surrounding electrons? Could we be sure that the advanced waves really would not cause any problems that contradicted experience? But everything seemed to work.

Electrodynamics Without Fields and the Principle of Least Action

In their investigations, Feynman and Wheeler discovered a remarkable property of their theory: they were able to build it on a principle of least action, where the action depended only on the paths of all electrons and no longer contained any electromagnetic fields. The paths of the electrons would then adjust themselves in such a way that this action would become minimal.

This was exactly what Feynman had had in mind from the very beginning. They had wanted the electromagnetic field to disappear as a mediator of forces. And if retarded and advanced fields are used on an equal footing, one can achieve a situation where the fields finally drop out completely and only time-delayed interactions between charges are actually needed.

The paths of the charged particles are very closely connected to each other, because the position of a charge at a certain time influences the other charges at times which lie both in its future and in its past, with the influence propagating both forward and backward in time at the speed of light. These charges in turn exert an influence on all other charges, and so on, so that a very dense network of mutual influences is created (Fig. 2.8).

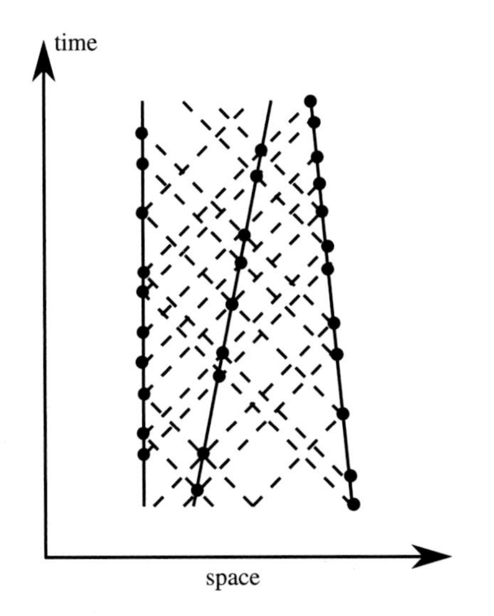

Fig. 2.8 Interaction network between three electrons (solid lines) according to the theory of Feynman and Wheeler. The interactions propagate at the speed of light into the future and the past (dashed diagonal lines)

Obviously, electrodynamics can be formulated in many different ways. We can use fields as mediators for the forces between charges, but we can also dispense with the fields and place a condition on the paths of the charges using a principle of least action. This provides a nice illustration of how helpful Feynman's view of reality in physics can sometimes be: instead of asking whether electromagnetic fields are real, he asks how useful they are, and comes up with different results depending on the formulation. In one formulation, they are indispensable and in a certain sense real. In the other formulation, they do not even exist. The question of whether fields are real is therefore beside the point because it leads us nowhere.

It is amazing that such a fundamental subject as electrodynamics can be expressed mathematically in so many different ways. This also happens in mechanics when we think of Newton's laws, Lagrange's formulation, or the principle of least action. Modern theories such as string theory make a living from the fact that they have various so-called dual formulations, which are not obviously equivalent. What does that mean? "I don't know", said Feynman in his Nobel Prize speech, speculating that it could be a sign of simplicity: "Perhaps a thing is simple if you can describe it fully in several different ways without immediately knowing that you are describing the same thing." A fascinating thought, about which we have certainly not heard the last word!

Monster Minds – Face to Face with the Geniuses

In good time, Feynman and Wheeler's alternative formulation of electrodynamics was sufficiently developed to be presented to a wider audience. Wheeler asked Feynman to give a seminar to gain experience in presenting scientific results and arranged a meeting with Eugene Wigner to put it on the regular seminar schedule. It was Feynman's first seminar of this kind.

A few days before the seminar, Feynman met Wigner and learned that he had invited the famous astronomer Henry Norris Russell. And that was not all: John von Neumann would also be there. Von Neumann was a great mathematician and one of the founders of computer science who, like Einstein, worked at the nearby Institute of Advanced Study. Feynman was gradually getting nervous, but there was more to come: Wolfgang Pauli (Fig. 2.9) was invited, one of the most brilliant physicists of the day, who had made decisive contributions to quantum theory. And even the great Albert Einstein would be present.

By this time, Feynman had turned pale, but Wheeler reassured him that he would be there and would answer the questions if necessary. And when the day came, Feynman filled the blackboard with complicated formulas in

Fig. 2.9 Paul Dirac (1902–1984, left), Wolfgang Pauli (1900–1958, center) und Rudolf Peierls (1907–1994, right) at a conference in Birmingham (© Science Museum London/ Science and Society Picture Library)

order to have something to hold on. He was very nervous and his hands were shaking – surely these monster minds would put him through the wringer! But when the seminar began and he had to concentrate on physics, all nervousness disappeared. He didn't care who was in the room any more – only the physical ideas counted. Feynman kept this ability throughout his life, and it must surely have been one of the reasons why his lectures were so unique.

When Feynman had finished his talk and the time for questions began, Pauli was the first to express his concerns. He was known as a harsh critic. Einstein, on the other hand, was open to the new theory. Many years earlier, he had himself investigated the symmetry of time in electrodynamics. In a dispute with the Swiss mathematician and physicist Walter Ritz, he had argued that the fundamental laws of electrodynamics do not favor any direction of time. Ritz had been of a different opinion. In his view, one had to avoid advanced fields, and this would make electrodynamics one of the reasons for the fundamental difference between the past and the future. In a joint publication from 1909 entitled *Zum gegenwärtigen Stand des Strahlungsproblems* (On the Present Status of the Radiation Problem), they wrote:

> Ritz considers the restriction to the form of retarded potentials as one of the roots of the second law, while Einstein believes that irreversibility is exclusively due to reasons of probability.

The second law mentioned here is the second law of thermodynamics, which most physicists take to specify the direction of time. Since the retarded potentials produce fields and waves that only propagate into the future, Feynman and Wheeler's theory thus tended to support Einstein's views. However, Einstein also remarked in the seminar that it would probably be very difficult to make a corresponding theory for gravitational interaction. There, the curvature of space and time is taken as the mediator for these forces, and this is so deeply embedded in the theory that it would hardly be possible to remove it.

Feynman later regretted that he could not remember Pauli's concerns – at that time he was just relieved at not having to answer his questions. Perhaps this experienced quantum physicist had noticed some problems which Feynman would only notice later when he tried to translate his theory into a quantum version. At the end of the seminar, Pauli asked Feynman what Wheeler was going to say about the quantum theory when he gave his talk. Wheeler had intended to work out and present the quantum theory relatively quickly. Feynman had to admit that he didn't know because Wheeler hadn't yet told him anything about it.

"Oh?" said Pauli. "The man works and doesn't tell his assistant what he's doing on the quantum theory?" Then he came closer and said in a low, secretive voice: "Wheeler will never give that seminar." He was right, because Wheeler found no way to formulate a quantum version of their theory. And neither was Feynman ever able to solve this problem in a satisfactory way. However, his work on it gave him completely new insights into the structure of quantum theories, as we will see in the next section.

2.3 The Action in Quantum Mechanics

Feynman and Wheeler had actually succeeded in removing the infinite self-interaction of electrons from classical electrodynamics. However, they had paid a price for this: they had had to do without the electromagnetic field produced by all the charges. In return, they had obtained an action that took on a minimum for the real paths of the charges. So Feynman was able to complete the first step of his idea with Wheeler, and the second step was to make a quantum version of their theory. After all, Feynman's aim was to eliminate the infinities in quantum electrodynamics and thus solve the central problem of that theory.

However, this project proved difficult. After the experienced Wheeler had failed to formulate a quantum version of their theory, Feynman also tackled this problem. But he couldn't get any further – Pauli's doubts were apparently justified! The recipes known at the time for converting a classical theory into a quantum theory simply didn't work. For these recipes to work, it had to be possible to consider the dynamics of all the particles at the same time. In classical mechanics, for example, one can simultaneously calculate the kinetic and potential energies of all particles at one time, calculate the difference, and then calculate the sum (more precisely, the time integral) over all points in time to determine the action.

But this was not possible with the action of Wheeler and Feynman, because it combined the locations and speeds of the particles at different points in time. This was due to the fact that, because of the finite speed of light, each particle would affect all other particles only with a time delay (or with a time advance). So the positions of many particles had to be considered simultaneously at different times in the action. The standard version of quantum mechanics could not handle this.

Dirac's Brilliant Idea

Feynman tried many ideas, but none of them helped. Then a coincidence came to his aid. One evening in the spring of 1941, he went to a party at the Nassau tavern in Princeton, where he met the German physicist Herbert Jehle who was a little over ten years older than Feynman. Like many others, Jehle had fled the Nazi terror in Germany. They got into conversation and Feynman asked Jehle if he knew a way to build a quantum theory directly from an action, given that Feynman and Wheeler had already found an action for their theory.

Jehle did not know any such way – indeed, nobody knew of one at that time – but he came up with something else. In 1932, Paul Dirac had published a paper in the *Soviet Union Journal of Physics* entitled *The Lagrangian in Quantum Mechanics*, in which he showed the relationship between Lagrangian mechanics and quantum mechanics. Perhaps you will remember from Chap. 1 that, in classical mechanics, the Lagrangian is the difference between the kinetic and potential energies of a particle, and the action is the time integral of this Lagrangian along the trajectory of the particle. Lagrangian and action are thus closely related!

The next day they went to the Princeton Library together and Jehle showed Feynman Dirac's paper. It was about the way a quantum wave develops over a very short period of time, working by analogy with Huygens' principle. At some chosen initial time, tiny elementary waves would thus form at each point of the quantum wave and subsequently spread. A short time later the quantum wave would then be the sum of all these elementary waves.

The question was: how big is such an elementary wave at some place, if it was created shortly before at another place? How often does the elementary wave swing up and down in the short time span, and does it have a wave crest or a wave trough at the destination? Dirac found that the number of oscillations is related to the Lagrangian of a particle which moves from the origin of the elementary wave to its destination in the short period of time. If you want to see the specific formula, here it is. The number n of oscillations in the short time span $\mathrm{d}t$ is given by $n = \mathrm{d}t \cdot L/h$, where L is the Lagrangian and h the Planck constant. In other words, the larger the product of the Lagrangian and the time span relative to the Planck constant, the more often the elementary wave will oscillate.

However, Dirac expressed himself somewhat imprecisely in his paper. He wrote that the exact strength of the elementary wave was *analogous* to a certain formula – only, what did *analogous* mean? Did it mean *equal*, or something else? Feynman wanted to find out immediately. So he jumped to the blackboard and started to calculate. Relatively quickly he saw that he had to insert another constant A. If he chose this constant appropriately, he actually succeeded in deriving the Schrödinger equation of quantum mechanics. Now it was clear what Dirac meant: not equal, but *proportional*!

When Jehle saw this, his eyes almost popped out. Apparently, this young American physicist had just made a pioneering discovery at the blackboard, something Feynman himself probably didn't fully realize at the time. So Jehle took out a little notebook and rapidly copied down what was on the blackboard.

In 1946 – a good five years later – Feynman met Dirac at the Princeton bicentennial and told him about his discovery that the elementary waves were proportional to Dirac's formula. "Are they?" said Dirac. "Yes!" Feynman replied. "Oh, that's interesting!" Dirac commented and went his way.

Did the great physicist really not know? It's hard to say. Dirac was a very quiet man, but at the same time, he was also an exceptional scientist, and twentieth century physics owes him much. It was difficult to see behind the façade of such a reserved man, who preferred to think things through alone and would generally answer questions with a short *yes* or *no*.

Dirac's reticence could sometimes be quite irritating to his colleagues, as revealed by the following little anecdote. In the 1930s, the young Dirac travelled to Copenhagen for a research trip organised by Niels Bohr. The generally talkative Bohr soon began to wonder about his young guest and complained to Dirac's colleague Ernest Rutherford: "This Dirac never says anything!" Rutherford then told him a short story in which an unsatisfied customer complains about a parrot he has bought at a pet shop: "The parrot never says anything!" The manager apologizes: "Oh, please forgive me. You wanted a parrot that talks, and I gave you the parrot that thinks!"

And Dirac certainly could think. Feynman and Dirac were probably equal in genius, but very different in character. Robert Oppenheimer quotes Wigner with the following words in a letter from 1943: "He (Feynman) is a second Dirac, only this time human." However, their differences in character did not prevent Feynman from admiring Dirac, who was sixteen years older than him, for his remarkable skills. As a young student, Feynman had learned quantum mechanics from Dirac's book *The Principles of Quantum Mechanics* – Dirac's clearly structured approach is still highly valued even today. When Dirac said or wrote something, it was usually of great importance. In 1986, in his Dirac Memorial Lecture in honor of Dirac, who had died two years earlier, Feynman said: "When I was a young man, Dirac was my hero. He made a breakthrough, a new method of doing physics. He had the courage to simply guess at the form of an equation, the equation we now call the Dirac equation, and to try to interpret it afterwards." That was exactly the style that Feynman also preferred: try to guess the solution intuitively, and don't worry about the details until later.

Path Integrals: A New Approach to Quantum Theory

Thanks to Dirac, Feynman now had in hand a connection between the quantum wave and the classical Lagrangian. Could he use it establish a connection between quantum waves and the classical action? Well, to obtain the action we have to integrate the Lagrangian over time, i.e., we cannot not limit ourselves to a short time interval. We must also consider longer periods of time. In the days following his meeting with Jehle, Feynman thought about how this could be achieved and finally found the solution. If he joined many short time intervals together and moved from one point to the next point, then to another point, and so on, the number of crests and troughs of the individual elementary waves that connect the points would add up (Fig. 2.10). With an infinite number of infinitesimal time intervals, the sum of crests and troughs becomes a time integral of the Lagrangian along the path that connects all the points taken, and that is exactly the action S of this path.

The situation is now completely analogous to Fermat's principle in Sect. 1.2. Suppose a quantum wave starts at a point A and we want to

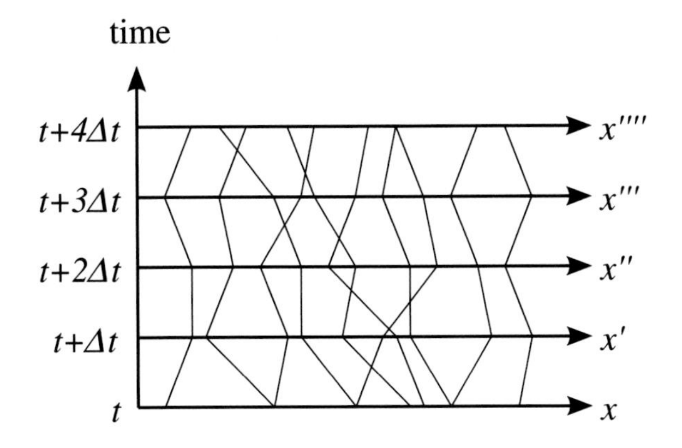

Fig. 2.10 How the path integral is formed in one space dimension

calculate what value it will have at a point B later. Then we can look at all the paths between points A and B. Each path provides a wave contribution to the value of the quantum wave at position B – in this sense, the quantum wave takes all possible paths between the two points. Whether a path at endpoint B contributes a wave crest or a wave trough to the total value of the wave depends on how many oscillations there are in this wave contribution. Feynman now knew that there would be n oscillations, where $n = S/h$.

In short, it is the classical action S of each path, expressed as a multiple of the Planck constant h, that determine whether this path will contribute a wave crest, a wave trough, or something in-between to the total wave at the end point. The sum of all these wave contributions along different paths – the so-called path integral – then determines the value of the wave function at the joint endpoint of the paths.

Actually we could almost have guessed this result, if one takes a closer look at Fermat's principle:

> If the possible paths that light can take are much longer than its wavelength and a continuum of paths is possible, the light chooses the fastest path between two points, i.e., the one it can travel in the shortest possible time.

This law follows from the fact that a light wave takes all possible paths between two points and that the individual wave contributions of each path add up at the endpoint, whence the light propagation time determines whether the path contributes a wave crest, a wave trough, or something in-between to the total wave at the endpoint.

In classical mechanics, Fermat's principle is replaced by the principle of least action: a particle selects the path with the least action between two points, in a situation where the start and arrival times are fixed. The light propagation time is therefore replaced by the action of the path. If we now assume that the principle of least action results from adding the wave contributions of many paths, the number of wave oscillations for each path must be proportional to its action. That was indeed what Feynman found.

However, Feynman's result goes further, because he was able to specify the exact number of wave oscillations for a given path: it is equal to the action, expressed as a multiple of the Planck constant. In other words, the wave swings up and down along a given path as often as the action of the path contains the Planck constant.

This also makes it clear when the description of a particle motion by one classical particle path is no longer sufficient, so that it has to be replaced by a quantum wave. Indeed, if the relevant paths are so short that their action is only the size of a few Planck constants, then all these paths are approximately equally important. This is the case, for example, for the motion of an electron in an atom. Every relevant path in an atom provides a wave contribution which contains only a few oscillations at most. And if there are only a few oscillations, wave effects will become visible!

However, if the paths are long and their action is much bigger than the Planck constant, the situation is different. The action for neighbouring paths usually differs by considerably more than a few Planck constants, so the paths contribute wave crests and troughs almost at random at the common endpoint and thus cancel each other to a large extent. Only those paths that are in the immediate vicinity of the path with the least action have very similar actions and thus similar wave amplitudes that can amplify each other. So the particle essentially takes the path with the least action and we observe a classical particle motion. This is essentially the same argument as we used to explain classical light beams by Fermat's principle in Sect. 1.2.

At this point I would like to add an additional remark for those readers who are already familiar with mathematics and physics, and especially with complex numbers (otherwise the following paragraph can be skipped without consequence for what follows).

Actually, it is not enough to talk about wave crests and troughs for quantum waves, because quantum waves are given by complex numbers at any place, and these must be imagined as arrows or clock hands in a two-dimensional plane, as discussed briefly in Chap. 1. Instead of a wave height, we have to specify arrow lengths and angles of rotation. Each path contributes such an arrow at the end point, and its angle of rotation is determined by the action of the path. The total quantum wave arrow at the end point is then the sum of all these arrows, where the arrows must all be joined together to form the sum. As can be seen in Fig. 2.11, it is the arrows of the path with the least action and its immediate neighboring paths that make the main contribution to the sum of all the arrows. With the more distant paths, the arrows turn faster and faster from path to path, the further away they are from the path with the least action, because the action of these paths grows faster and faster. In the sum, they no longer make a noteworthy contribution, because their joined arrows curl up in a spiral.

With his path integrals, Feynman had found a completely new approach to quantum mechanics, which was in many respects more elegant than the previously known formulations. Once the action of a classical theory was found, it should therefore be possible to formulate the corresponding

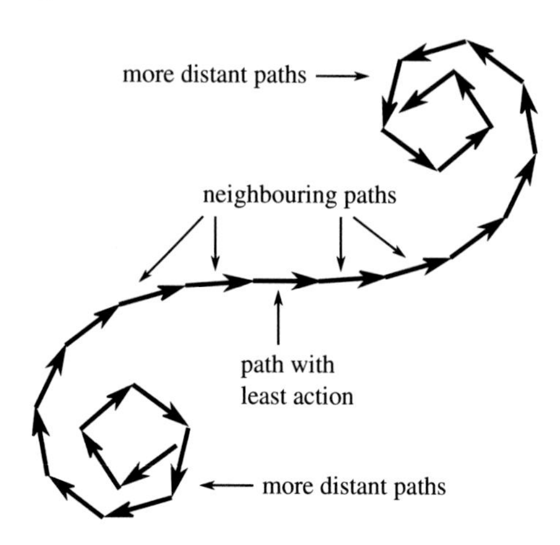

Fig. 2.11 Sum of the quantum wave arrows of the different paths at their common endpoint, as described in the text. Feynman repeatedly used such images to explain how the principle of least action results from quantum mechanics

quantum theory using path integrals. The derivation of the principle of least action of a classical theory from quantum theory was crystal clear – it was to be done just as Fermat's principle for light rays was derived from the wave description of light.

Feynman's formulation has yet another advantage that becomes important when the Special Theory of Relativity is included: the formulation is not based on differential equations like the Schrödinger equation, which describes the rate of change of a system at any time. Such differential equations assume that the system can be specified at a certain time, for example, by a quantum wave. In the Special Theory of Relativity, however, different observers consider different events to be simultaneous, depending on their relative motion. Each observer sees other "slices" of simultaneity in spacetime. Thus, if an observer specifies a quantum wave in space at a certain time, then this wave cannot be used for another observer, because it will comprise space points at different times for that observer. This leads to complications, because it is not immediately obvious how to convert the quantum wave between different observers so that it only includes space points at what the new observer considers to be a fixed time.

Feynman's formulation did not have this problem, because it was based on complete paths in space and time. It was easy to consider these paths in one or other different observer perspectives, so Feynman's path integrals were perfect for including the Special Theory of Relativity.

Something Is Wrong

Feynman was therefore confident that his method could solve the problems of quantum electrodynamics, for which the inclusion of special relativity was essential. Equipped with his new action for electrons without self-interaction, he set to work – and again encountered problems. For example, it was difficult to describe the spin of the electrons, i.e., their quantum angular momentum. In addition, the fact that Feynman's action took several points in time into account simultaneously meant that it was no longer possible to work with quantum wave functions. The entire system of interacting electrons could not be described by a wave function at some given time.

But Feynman did in fact find a solution. The whole thing had to be handled as a scattering process. Quantum mechanical scattering processes were Wheeler's area of expertise, and Feynman had learned from him that practically every quantum mechanical problem can be treated as a scattering process.

In such a process, energetic particles are generally fired at a system – for example an atom – and we measure the directions in which they are scattered. Fast particles can also be fired directly at each other. At the Large Hadron Collider (LHC), high-energy protons collide with each other, and if for example two high-energy photons are generated, we can determine what energy they have and in which direction they leave the collision zone. In this way, the famous Higgs particle was discovered in 2012 (Fig. 2.12): If a short-lived Higgs particle is created in the collision and decays directly into two photons, this pair of photons has a characteristic total energy that corresponds to the mass of the Higgs particle.

Fig. 2.12 Collision event in the CMS detector at the LHC, showing two high-energy photons (green lines to the top and bottom right) emitted in the decay of a Higgs particle that had just formed (© CERN, CMS-Experiment, see http://cms.web.cern.ch/news/observation-new-particle-mass-125-gev)

The advantage of dealing with scattering processes is that we do not have to worry about the exact sequence of events, so we can do without a wave function that represents the time development of the process. It suffices to specify quantum-mechanical probabilities for the experiment as a whole, i.e., for example, the probability that two photons are emitted in certain directions during a proton collision at the LHC.

At times, it really looked as if all the problems could be solved by looking at scattering processes. But then Feynman got a kind of funny feeling that things weren't exactly right, as he puts it in his Noble Prize speech.

In 1942, Feynman published his doctoral thesis entitled *The Principle of Least Action in Quantum Mechanics*, in which he presented his new approach to quantum mechanics with the path integral. The application to quantum electrodynamics was not included, because he had discovered in the meantime that certain things were going wrong – in fact, certain things were even fatal!

For example, the quantum theory with Feynman and Wheeler's action yielded energy values that had to be described by complex numbers. But even in modern physics, energies have to be ordinary real numbers. Complex energy values are a sign of an instability in the theory. Furthermore, Feynman found that the probabilities did not add up to unity. So the total probability of something happening, no matter what it might be, was not unity, and this can never be true! Apparently, Wolfgang Pauli had been quite right in the objections he had expressed in the seminar.

After the war, Feynman would once again address these problems and develop his own approach to quantum electrodynamics. And while he would never succeed in saving the approach devised with Wheeler, his semi-intuitive path integral method would eventually prove extremely useful.

At this particular time, however, other things had become more urgent. America was at war with Germany and Japan. In addition, the feasibility of an atomic bomb had been much discussed and the big question was: what if Nazi Germany were the first to develop and use such a bomb? This had to be prevented at all costs. So America launched a major military research project in which Feynman was to play a role.

2.4 Radioactivity and the Manhattan Project

There are many good reasons for calling the first half of the 20th century the age of nuclear physics. Even today, nuclear physics is still surrounded by an aura of mystery, and it has left its mark in the consciousness of mankind,

Fig. 2.13 Pierre and Marie Curie in their laboratory around 1898 (© The Print Collector/picture alliance/Heritage Images)

not least because of the atomic bomb and the disasters at Chernobyl and Fukushima.

It all started in a quite harmless way. In 1896, the French physicist Antoine Henri Becquerel discovered that uranium emits a special form of radiation known today as radioactive radiation. Like the X-rays discovered shortly before, this radiation caused a blackening of photographic plates, and it could penetrate opaque matter and ionize air. Slightly later, the physicists Marie and Pierre Curie (Fig. 2.13) discovered that the element thorium was also radioactive. They also identified two new elements, which were even more radioactive than uranium or thorium, calling them radium and polonium (as Marie Curie came from Poland).

Over the next few years, much progress was made in understanding the nature of this radiation. There were three different types of radiation with very different characteristics: first, alpha radiation, which consists of positively charged alpha particles (i.e., helium nuclei), then beta radiation, which is made up of high-energy electrons (and sometimes positrons), and finally the penetrating gamma radiation, which resembles X-rays, but is in fact a more energetic electromagnetic wave, with a shorter wavelength.

Radioactivity: Atomic Nuclei in the Quest for Stability

In 1910, Ernest Rutherford and his doctoral student Ernest Marsden found the origin of this radiation with their famous scattering experiment. The radioactive radiation comes from the atomic nucleus, a tiny positively charged entity at the center of each atom, ten thousand times smaller than the atom itself. This atomic nucleus contains almost the entire mass of the atom and is surrounded by the light, negatively charged electrons of the atomic shell.

Many other experiments were required to determine the structure of atomic nuclei (Fig. 2.14). These consist of two different types of particle: positively charged protons and neutral neutrons. A very strong force – the so-called nuclear force or strong interaction – holds these particles together against the electrical repulsion of the protons in the nucleus. The nuclear force has a very short range, acting only between directly adjacent protons or neutrons, and it makes no distinction between these particles. Protons and neutrons are held together by the nuclear force in the atomic nucleus rather like the molecules in a drop of water. Atomic nuclei can therefore be thought of as tiny liquid droplets, especially if they contain many protons and neutrons.

If atomic nuclei contain a large number of protons, the electrical repulsion between them increases. Each proton feels the electrical repulsion of all the other protons in the nucleus, but is only held by the strong force to its

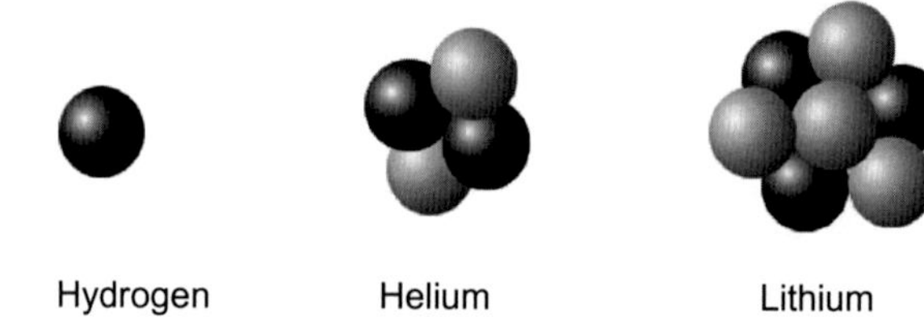

Fig. 2.14 The most common atomic nuclei of the first three elements in the periodic table. Protons are shown as dark spheres, neutrons as light spheres. Hydrogen can also occur with one or two neutrons and helium with only one neutron, but these so-called isotopes are much rarer

immediate neighbors in the nucleus. Beyond a certain number of protons, the problem of stability becomes more delicate, until the electrical repulsion becomes too great. The nucleus becomes unstable and tries to rid itself of protons. However, it does not usually kick them out individually, emitting in preference a particularly stable package of two protons and two neutrons, i.e., an alpha particle (helium nucleus). The largest stable atomic nucleus belongs to lead, with 82 protons and between 124 and 126 neutrons. As we see, the number of neutrons in an element can vary, because the number of protons alone determines which chemical element it is. One speaks of different isotopes of an element, in this case lead-206, -207, and -208, where the attached number is the total number of nuclear particles (nucleons), i.e., protons and neutrons.

The origin of alpha radiation is thus clear: large atomic nuclei are ridding themselves of excess protons. But the situation is different with beta radiation, which originates in the transformation of neutrons into protons, whereby a high-energy electron and a so-called neutrino are formed and emitted by the nucleus (Fig. 2.15). This is advantageous for atomic nuclei with many neutrons, because a neutron is a little bit heavier than a proton, and when it transforms into a proton plus electron plus neutrino, some mass is converted into energy and released, thereby stabilizing the nucleus.

Unlike alpha and beta radiation, gamma radiation does not change the number of protons and neutrons in an atomic nucleus. It arises when an energetically excited atomic nucleus, which can be imagined as a vibrating drop of water, radiates this excitation energy in the form of high-energy photons. This is very similar to what happens with excited electrons in the shell of an atom, which radiate excitation energy as visible light, with the difference that photons of gamma radiation are millions of times more energetic than photons of visible light.

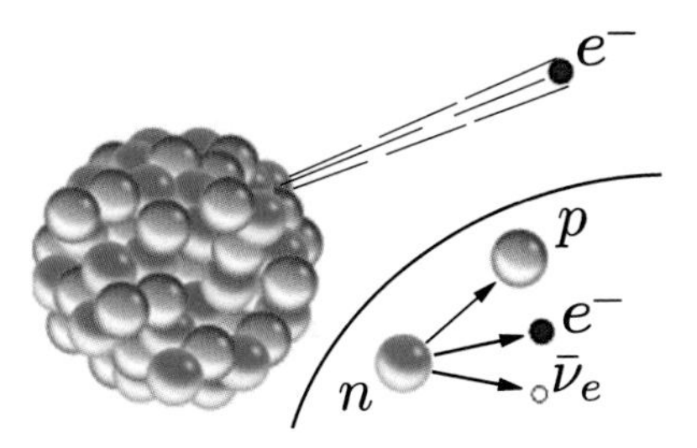

Fig. 2.15 In beta decay, a neutron (n) converts into a proton (p), leading to emission of a high-energy electron e^- and a neutrino. Source: https://commons.wikimedia.org/wiki/File:Beta-minus_Decay.svg?uselang=de

Where do these photons come from, when emitted by an atom as light or by an atomic nucleus as gamma radiation? And where do the electrons come from, when emitted by the transformation of a neutron into a proton and perceived as radioactive beta radiation? Does the photon or electron already exist in the atom or atomic nucleus before? This is exactly the question that Richard's father Melville once asked his son, who was now a student, when he visited his parents at home. Melville had always been very interested in science, as we know, even though he had never had the opportunity to become a scientist himself. Now that his son had become a physicist, he hoped he would finally get a clear explanation.

But Feynman found it difficult to give his father a satisfactory answer.[5] In fact, these photons and electrons are created from the energy available. They are not actually there before, but how can we picture that? How can we view this idea that a particle like the photon can come out without having been there before?

After a few moments of thought, Feynman gave up: "I'm sorry; I don't know. I can't explain it to you." Melville was quite disappointed – so many years of study to get such a poor answer to such a simple question. But quantum physics has nothing more to offer! It does not provide any detailed mechanism to explain how a photon or electron is created during a decay process. All that matters is that the energy is sufficient for the formation of the given particle, that the laws of conservation for charges, momentum, and so on, are respected, and that the probability for the event can be calculated. It may be disappointing, but according to today's understanding, quantum physics can do no more than that. Even a teacher as experienced as Richard Feynman could not change that.

Some Atomic Nuclei Can Be Split

In the first three decades of the twentieth century, nuclear physics was still regarded as a somewhat exotic "pure" science, far removed from any serious application. However, this began to change in 1938, when Otto Hahn (Fig. 2.16) and his assistant Fritz Straßmann at the Berlin Kaiser Wilhelm Institute for Chemistry demonstrated a completely new kind of nuclear reaction. They had exposed uranium to neutron radiation, expecting these

[5]See *What is Science?* http://www.fotuva.org/feynman/what_is_science.html (taken from *The Physics Teacher*, 1969).

Fig. 2.16 Lise Meitner and Otto Hahn in the laboratory at the Kaiser Wilhelm Institute of Chemistry (1913) (© picture alliance/Everett Collection)

neutrons to be captured by the uranium nuclei or get stuck inside them. The idea was that they would subsequently be transformed into protons by beta decay, thereby forming new, heavier elements with more protons, the so-called transuranium elements.

It is easy to imagine how surprised they were when they detected the mid-heavy elements barium and krypton instead. Where did these elements come from? A look at the proton numbers gives the answer: barium has 56 protons and krypton 36, so together they make up the 92 protons of a uranium nucleus. Apparently, uranium nuclei had burst into two parts when they were hit by neutrons. The neutrons had split uranium nuclei. No one had expected that!

Shortly after the discovery, the physicist Lise Meitner (Fig. 2.16), who had already been expelled from Germany at that time, and her nephew Otto Frisch suggested the following way to understand this nuclear fission. With its 92 protons, uranium nuclei already constitute a relatively unsta-

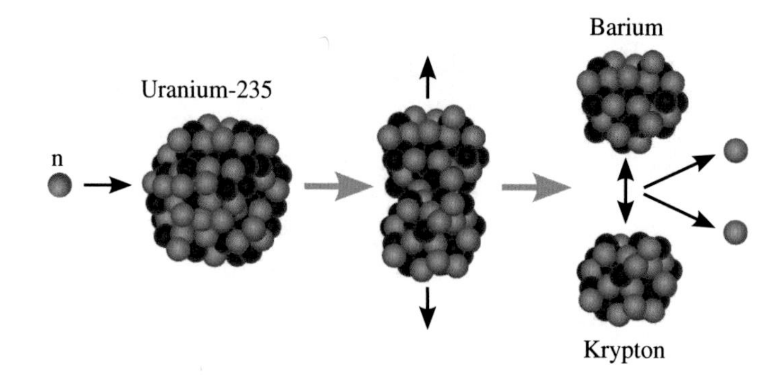

Fig. 2.17 Nuclear fission of uranium

ble structure, only held together, by the nuclear force, with great difficulty against the electrical repulsion of the 92 protons. If a uranium nucleus catches a passing neutron, it is set into vibration like a drop of water. During the resulting oscillations, the electrical repulsion of the protons can temporarily become stronger than the nuclear force and tear the core into two parts (Fig. 2.17). The two fragments then fly apart with high energy, driven by their very strong electrical repulsion. In this way, millions of times more energy is released than in chemical reactions, since the forces acting in the nucleus are much greater than the forces in the atomic shell that are relevant for chemistry.

If a uranium nucleus is split by a neutron, two to three new neutrons are released. These can in turn trigger further nuclear fission, releasing additional neutrons, and so on. So if we can accumulate enough fissionable uranium, so that more neutrons are created than are consumed, a devastating chain reaction can start, releasing far more energy in a fraction of a second than any chemical bomb.

Following the discovery of nuclear fission, everyone soon became aware of the enormous threat posed by such a nuclear atomic bomb. Would Nazi Germany be able to build this bomb? It seemed quite possible – after all, nuclear fission had been discovered in Berlin!

Some physicists in America were very concerned and saw the need to inform their government of the possible danger. In July 1939, the Hungarian-born physicist Leó Szilárd, together with Edward Teller and Eugene Wigner, wrote a letter to the American President Franklin D Roosevelt on behalf of Albert Einstein. Einstein had at first been surprised when he had heard about the possibility of a nuclear chain reaction – he hadn't yet thought of such a thing! If the Nazis could get their hands on such a devastating bomb, that would be fatal. Einstein himself had had a bad experience with the Nazis because of his Jewish descent and liberal mentality – he had understood the criminal intentions of the Nazis earlier than many others and had already turned his back on Germany in 1933.

So the otherwise pacifist Einstein signed the letter.[6] When it became clear after the war that the danger of a German atomic bomb had been overestimated, Einstein regretted his decision: "Had I known that the Germans would not succeed in developing an atomic bomb, I would have done nothing."

America Builds the Bomb

The United States was determined to pre-empt the Germans in building an atomic bomb and hence launched the Manhattan project, which would eventually include most of the physicists in the United States. Other scientific activities in the USA almost came to a standstill as nearly all physicists tried to make their contribution to win the alleged race.

In 1942 Feynman was also approached in Princeton. Robert (Bob) Wilson – then an experimental physicist at Princeton – visited Feynman in his office and asked him if he wanted to work on a highly secret project to build an atomic bomb. At first, Feynman wasn't very enthusiastic, but Wilson ignored his rejection: "We have a meeting at three o'clock. I'll see you there!"

Wilson was right – Feynman could not resist such a challenge for long. The Germans had elected Hitler and the possibility of developing an atomic bomb was obvious. The chance that the Germans would develop such a bomb before the Americans was terrifying. So Feynman went to the meeting. By four o'clock he already had a desk in a room and was trying to figure out whether Wilson's idea could really work.

So what was the problem? When we said above that a neutron can split a uranium atom, that was only half the truth. Natural uranium is found basically in two variants or isotopes: uranium-235 and uranium-238 with three more neutrons – the proton number is 92 in both cases, otherwise it would not be uranium.

[6] The letter and Einstein's comments on it can be found on Wikipedia under the heading *Einstein–Szilárd letter.*

The three additional neutrons in uranium-238 stabilize the atomic nucleus, as they allow the positively charged protons to move a little further apart, so that their electrical repulsion becomes weaker. This means that this isotope has a long half-life, some 4.5 billion years, which happens to be roughly equal to the age of the Earth. Hence, about half of all uranium-238 atoms found on Earth at the time of its formation have by now disintegrated.

Uranium-235 is significantly less stable. It has a half-life of around 704 million years, about six times shorter than uranium-238, which means that, during the existence of the Earth, the amount of uranium-235 on it has been halved six times and now amounts to only 1/64, or slightly more than 1% of the original amount. So it turns out that approximately 99.3% of the uranium found in nature is uranium-238 and only 0.7% is uranium-235.

But it is the more unstable uranium-235 nucleus that can be split by neutrons – uranium-238 is too stable for this. To build an atomic bomb, the amount of fissile uranium-235 must first be artificially increased, i.e., the uranium must be enriched. Chemical methods cannot be used for this because the two uranium isotopes have the same electron configuration and therefore react the same way chemically.

Thus, the isotopes must be physically separated by exploiting the fact that they have slightly different masses. Fortunately, given the potential consequences, this is not easy. It requires a great deal of technical effort. In principle, several different methods are available, one of which was developed by Wilson.

Feynman was completely taken up by the new project. As mentioned above, nearly all physicists in the USA had stopped their other research and were trying to contribute in some way with their knowledge and technical resources. Feynman had also postponed the completion of his thesis, and had to take six weeks off from the project in between in order to complete it – Wheeler had insisted.

Marriage with Arline

As soon as he had his degree in hand, Feynman kept his promise to Arline and arranged their wedding, against the will of both their parents. Due to Arline's severe tuberculosis disease, they saw no future for the young couple and were afraid that Richard could also be infected – tuberculosis was a death sentence at that time, when antibiotic treatments were practically non-existent. But Richard and Arline had a different opinion and refused to view the future with such pessimism. They would have a happy time together as a married couple, no matter how short this time might be!

On their wedding day, Richard borrowed a station wagon and put in a mattress so that Arline could rest during the trip. Then he picked her up and they drove to their lonely wedding ceremony, in which neither parents nor friends took part. Arline was already too sick for a honeymoon, so Richard took his young wife straight to a hospital near Princeton in New Jersey. It's heart-breaking to see how tough Richard and Arline had to be to face this destiny. Because of the risk of infection, even a kiss was dangerous and an exhausting pregnancy would have been a medical disaster for Arline. Nevertheless, it seems a shame that the couple had not received more support from their parents. There can be little doubt that they had underestimated the importance of the extraordinary bond between Richard and Arline. It would take a while before Richard's relationship with his mother in particular would return to normal – she had spoken out particularly vehemently against the marriage through concern for her son, but after Arline's death she recognized her mistake and sought reconciliation.

When Feynman returned to Princeton after the wedding, it was decided not to proceed with Wilson's separation process for the uranium isotopes, since other methods were more promising. For Feynman, the decision was understandable. He admired the efficiency with which men like Compton, Tolman, Smyth, Urey, Rabi, and Oppenheimer weighed up the various arguments and then came quickly to a rational decision. "These were very great men indeed", he writes in *Los Alamos From Below*, and he was pleased that the project gave him the opportunity to meet and work with such men. This was exactly the benefit that Feynman would gain from the Manhattan project: he had the opportunity to work with the most influential physicists of his time, and they also had the opportunity to recognize the qualities of the young Feynman.

Going to Los Alamos

In the end, it was decided to start the project to build the bomb in a small and remote village called *Los Alamos* in New Mexico, near Santa Fe. As the buildings in Los Alamos were not finished yet, Wilson sent Feynman to work with Enrico Fermi and Wheeler in Chicago, where many contributions to the project were being undertaken at that time. When completion of the laboratories in Los Alamos was further delayed, the project members in Princeton became impatient and decided to leave for Los Alamos any-

Fig. 2.18 Hans Bethe (1906–2005, left) and Robert Oppenheimer (1904–1967, right) (© picturealliance/akg images)

way, because there was nothing more they could do in Princeton, and they thought that maybe the construction could be speeded up once they arrived. Robert Oppenheimer (Fig. 2.18 right) was the scientific director and proved to be a very good choice. He was attentive to the needs of those working with him, and every member of his team, including Feynman, felt great respect for him. He also took care of Arline's hospital requirements in Albuquerque, about 100 miles away, where Feynman visited her as often as possible.

The buildings and laboratories were eventually completed and the project gained momentum. Feynman was assigned to Hans Bethe's theory group (Fig. 2.18 left) and they developed a close friendship. They complemented each other well. Bethe, who was 12 years older and whose pronounced German accent can be heard on many videos on the Internet, was quiet and calm, in contrast to Feynman, who reacted rather impulsively and spontaneously. Like Einstein, Bethe had fled from the Nazis early on, first emigrating to Great Britain in 1933, and later to the USA. In the 1930s, he was one of the leading nuclear physicists, and this made him particularly valuable for the project. In 1939, for example, he explained how the Sun and other stars use nuclear fusion to generate their energy.

Bethe and Feynman often got into intense physical discussions in which Feynman, as always, completely forgot what an outstanding physicist he was dealing with: "No, no, you're crazy. It'll go like this", he would shout for example. But Bethe did not resent him because this was exactly the kind of intense discussion he was looking for. Soon their colleagues knew the game and coined the expression: "The battleship and the mosquito boat", where Feynman was clearly not the battleship. Feynman also enjoyed these discussions, and he admired Bethe for his ingenuity and inexhaustible collection of useful mathematical tricks. Bethe also knew that Feynman was an outstanding talent in his team, and Feynman soon became the leader of a four-man group.

Other physicists also sought this kind of straightforward discussion with Feynman. When Niels Bohr and his son Aage came to Los Alamos from Denmark one day, there were important meetings to discuss problems with the construction of the bomb. Everyone wanted to see the famous man, and the meetings were just packed, so Feynman hardly got to see him.

The next morning Feynman received a phone call: Niels and Aage Bohr wanted to talk to him alone. Feynman was surprised, because at that time he was a young and relatively unknown physicist. Nevertheless, Niels Bohr had noticed him. So Feynman joined the two Danish physicists and they began to discuss the design of the bomb, without disturbance from others. The Bohrs had some ideas on how to make the bomb more effective, and they were looking for someone who would not freeze in awe of them, but who would honestly and openly express their opinion. That was exactly the right thing for Feynman, because he never minced his words when it came to physics – no matter who he was talking to! Even in later years, Feynman maintained this habit, although he occasionally exaggerated somewhat.

Another star of the physics community was the Italian-born Enrico Fermi (Fig. 2.19). Once when he came down from Chicago, Feynman discussed with him the result of an elaborate calculation which he had not fully understood. Fermi thought about it and soon came up with a clear explanation for the result, which Feynman hadn't thought of, although he had been reflecting on the problem for some time. "He was doing what I was supposed to be good at, ten times better", Feynman writes in *Los Alamos From Below*.

John von Neumann, the great mathematician, was also in Los Alamos. Feynman liked to go for a walk with him on Sundays, and Bethe often joined them. Feynman enjoyed these encounters. And von Neumann gave him an interesting idea: "You don't have to be responsible for the world that you're in." This sentence would have a liberating effect on Feynman in his later life – it's easy to make ourselves unhappy if we take too many things to heart and consider ourselves responsible for everything: "It's made me a very happy man ever since", as Feynman himself put it.

You Need to Know What You're Doing

The time in Los Alamos must have been exhausting for Feynman, even though he often overplayed this in his own books and lectures by telling

Fig. 2.19 Colloquium at Los Alamos 1946. Front row left to right: Norris Bradbury, John Manley, Enrico Fermi, and J. M. B. Kellogg. Second row left to right: Colonel Oliver G. Haywood, unknown, Robert Oppenheimer, Richard Feynman, and Phil B. Porter. Source: Los Alamos National Laboratory

so many amusing stories. They were under pressure for time, because they wanted to get results before the Germans and there was a lot to learn. Many problems had to be solved, involving the collaboration of many different people, and all this had to be carefully organized. For example, how could they calculate the explosive strength of an atomic bomb? With Bethe, Feynman succeeded in creating a formula that became known as the Bethe – Feynman formula – presumably, certain aspects of this formula are still kept secret even today.

One of Feynman's tasks was to organize and perform extensive numerical calculations. Today this would be done with the help of modern computers, but at that time they were still dependent on mechanical gadgets, although later they had IBM machines into which cards were inserted, which could perform easy tasks like multiplying or adding two numbers. Technical and organizational problems came up again and again, but here Feynman was in his element – he liked computers, and he loved getting to the bottom of problems, never giving up until he had found a solution.

In doing so, Feynman insisted that it was necessary for efficient cooperation to get people fully involved and make it clear to them why they were doing whatever it was. But for reasons of secrecy, the US army pursued exactly the opposite approach: everyone should only know what they absolutely had to know to carry out their given task! In Feynman's

Fig. 2.20 Aerial view of the K-25 gaseous diffusion plant for uranium enrichment in Oak Ridge. Source: American Museum of Science and Energy, James E. Westcott, official US Army Manhattan Project photographer

view, this could not work efficiently, so he told the group which carried out the numerical calculations and had to struggle with stacks of punch cards every day exactly what it was their work was needed for: "We're fighting a war!" This changed the situation completely! The group began to invent ways of doing things better and speeding up the numerical calculations.

A similar situation occurred at the large Oak Ridge facility (Fig. 2.20) in Tennessee, where they were trying to separate the isotopes of uranium. Here, too, those involved did not know what their work was really needed for. They had not been informed about the physical background, but according to army policy, were blindly following the instructions they had received. The results were correspondingly poor – only a small amount of uranium-235 had actually been concentrated. They had also stored large quantities of uranium nitrate solution in tanks and had no idea of the dangers involved. When Emilio Segrè, who had received his Ph.D. from Enrico Fermi and fled from the Italian fascists in 1938, travelled to Oak Ridge, he was literally appalled. If they later dealt with enriched uranium in the same way, the amounts could be sufficient to trigger a significant chain reaction. There might be an explosion or at least a contamination of the plant with radioactive fission products. When Segrè drew the attention of the officials to the danger, they were understandably alarmed. An explosion?

Hadn't the Army taken that into account when they built the facility? Well, the army knew that about 20 kilograms of enriched uranium would be needed to make a nuclear bomb, and they also knew that there would never be this much in the plant at any time, so they were not worried. But they had ignored one very important fact: neutrons are considerably more effective when they are slowed down in water. This meant that a uranium nitrate solution would require less than a tenth or even a hundredth as much material to trigger a nuclear chain reaction.

In Los Alamos, the necessary calculations were made as fast as possible – how much uranium salt solution and how much dried uranium salt could be stored safely in one place? Feynman was eventually sent to Oak Ridge to present the results and explain the corresponding physics. He was a little concerned that no one would listen to a young physicist like him, so Oppenheimer gave him a formulation that would work wonders: "Los Alamos cannot accept the responsibility for the safety of the Oak Ridge plant unless … ." And so we see how important a little psychology can be in such situations!

Feynman explained to the managers and technical experts in Oak Ridge why the uranium was being enriched, how effectively fast and slow neutrons could trigger a chain reaction, how the uranium could be shielded, and so on. And so it was made clear to the people in the plant what mistakes they had unwittingly being making, and what steps had to be taken in the future.

Annoying Censors and Cracking Safes

With all these challenges and his constant concern for his sick wife, it was clear that Feynman needed some kind of compensation. He found it, for example, by engaging in battles with the army censors, who controlled and censored all incoming and outgoing letters at Los Alamos. This was rather annoying and even illegal, and was therefore carried out on a "voluntary basis".

The censors were particularly concerned when they suspected secret codes in the letters. Richard once wrote to Arline, who was bored in her clinic, about an observation he had made while playing around with the calculators in Los Alamos: if you divide 1 by 243, you get a decimal number in which the 27 digits 411522633744855967078189300 always repeat[7]:

$$1/243 = 0.004115226337448559670781893004115226337448 55\ldots$$

---

[7] If you want to calculate it yourself, see, e.g., the website http://www.wolframalpha.com/.

Feynman found this amusing and also somewhat strange: first 411, then 522 and 633, finally 744 and 855, before it gets a bit chaotic after 96 and then starts again from the beginning. However, the censors suspected a secret message in this sequence of digits, and it took Feynman some effort to convince them that he had only written down the result of 1 divided by 243.

Feynman also enjoyed decoding the security combinations of the locks on his colleagues' lockers and safes, where they kept their secret documents, and soon gained a reputation as a safecracker. In *Los Alamos From Below* he describes these anecdotes in a very humorous way, giving the impression that the whole project was actually a great deal of fun for him.

Farewell to Arline

But then reality caught up with him in a rather brutal way: Arline's health was getting worse and worse, and all Feynman's efforts to find a successful treatment for her illness were in vain. In mid-June 1945, he borrowed the car from his friend Klaus Fuchs, who was later unmasked as a German spy, and drove to the clinic in Albuquerque to be with her in her last hour. She fell asleep gently and her breathing became less and less perceptible until no more breath could be felt. She was only 25 years old. Feynman took her belongings, took care of the paperwork, and drove back to the office as if nothing had happened. Deep down inside though, Arline's death must have hit him hard: "I was supposed to be supervising everything, but I couldn't do it for three days." A month later he was struck by grief when he saw a beautiful dress in a shop window in Oak Ridge. Arline might like that, he thought, and then he realized she would never be able to wear another dress.

How deep and long-lasting his grief for Arline must have been becomes clear from a letter he wrote to her on 17 October 1946, 16 months after her death. He even put it in an envelope and closed it, but it wasn't until after his death that the letter was found in his estate. It's a very private letter, and so moving that one hardly dares to read it. Nowadays it can be found on the Internet, for example at http://www.lettersofnote.com. We have reproduced some short excerpts from it in Infobox 2.3, because it shows how much Richard loved Arline, even after many months, and how alone he felt without her. No other woman had any chance of filling the gap that her death had left. "You, dead, are so much better than anyone else alive," Feynman writes, and concludes with the helpless words: "PS Please excuse my not mailing this – but I don't know your new address."

Infobox 2.3: Feynman's letter to his deceased wife Arline

October 17, 1946

D'Arline,

[...]

It is such a terribly long time since I last wrote to you – almost two years but I know you'll excuse me because you understand how I am, stubborn and realistic; and I thought there was no sense to writing.

[...]

I find it hard to understand in my mind what it means to love you after you are dead – but I still want to comfort and take care of you – and I want you to love me and care for me. [...] Can't I do something now? No. I am alone without you and you were the "idea-woman" and general instigator of all our wild adventures.

[...] you can give me nothing now yet I love you so that you stand in my way of loving anyone else – but I want you to stand there. You, dead, are so much better than anyone else alive.

[...] I have met many girls and very nice ones and I don't want to remain alone – but in two or three meetings they all seem ashes. You only are left to me. You are real.

My darling wife, I do adore you.

I love my wife. My wife is dead.

Rich.

PS Please excuse my not mailing this – but I don't know your new address.

The Bomb Explodes

The sensitive Bethe noticed that Feynman desperately needed a break after Arline's death and sent him home to his parents in Long Island. About four weeks later he received an encrypted telegram there, which called him back to Los Alamos. On 15 July he arrived there, and shortly afterwards he and his colleagues were driven to a remote desert area where the first test of an atomic bomb was to take place in the morning hours of 16 July 1945 – the Trinity test (Fig. 2.21).

Feynman belonged to a group that was stationed 20 miles (32 km) away from the blast site. All received dark glasses to protect their eyes from the bright light, but Feynman decided that a windshield would provide enough protection from the UV rays. So he was one of the few – perhaps even the only one – to see the explosion with his naked eyes.

The bomb was actually made of fissile plutonium and not uranium. It exploded with an explosive force of around twenty thousand tons of TNT. The flash of light was so bright that Feynman ducked and saw a purple

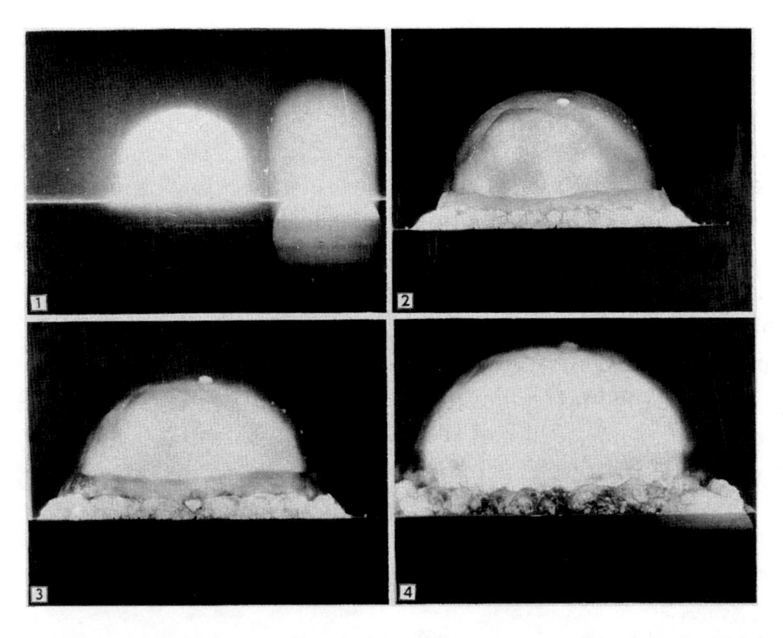

Fig. 2.21 Sequence of photos provided by the U.S. Army showing the Trinity test from a distance of about 10 km (© picture alliance/AP Images)

splotch as an after-image on the floor. Then he saw how the bright white light slowly faded and changed to yellow, then orange. Clouds formed out of nowhere and disappeared again due to the passing pressure wave. Finally, a large glowing orange ball rose up and formed the well-known atomic mushroom cloud.

Everyone stared quietly at the glowing spectacle on the horizon. Feynman was fascinated by the physical phenomena caused by the massive explosion. After about a minute and a half they were hit by the sound and pressure wave. There was a sharp BANG and then a rumble, like thunder.

This sound suddenly released everybody. So it had actually worked! All their efforts had finally been rewarded. It is easy to imagine what happened afterwards in Los Alamos. There was tremendous excitement. They had parties, drank, danced, and laughed a lot. But there was one man who was just sitting there moping. It was Bob Wilson, who had persuaded Feynman to help build the bomb three years earlier, and who was now saying miserably: "It's a terrible thing that we made!" It was only then that Feynman realized what had happened: they had started to build the bomb for a good reason, and they had worked very hard to make the project a success. It had been exhausting and exciting, and they had stopped thinking. "And you stop thinking, you know; you just stop," writes Feynman.

On August 6, 1945, the first atomic bomb was dropped on the Japanese city of Hiroshima, and three days later the second atomic bomb was dropped on Nagasaki. These devastating events must have triggered something in Feynman's mind. He describes how he sat in a restaurant in New York some time later and imagined what damage the Hiroshima bomb would do in New York. All the buildings he saw would be smashed. Then he saw people who were building a bridge or making a new road and he thought: "They are crazy, they just don't understand. Why are they making new things? It's so useless."

Feynman believed that a devastating nuclear war was almost inevitable sooner or later. The impressions of the recent war and the terrible impact of the atomic bombs in Japan had had their effect. Why build anything? Why go on exploring the fundamental laws of nature? This sense of futility was to accompany Feynman for many months after he had left Los Alamos in October 1945 and returned to his old life as a university scientist.

3

Feynman's Path to Quantum Electrodynamics

The end of the Manhattan project marked the beginning of a new phase in Feynman's life. He followed Hans Bethe to Cornell University and struggled at first to get used to his new life as a university professor – the dramatic experiences of the last few months had been too deep. But he gradually overcame his depression and began to blossom again. He worked tirelessly on the semi-intuitive approach to quantum electrodynamics (QED) that his path integrals had opened up for him. Eventually, this effort was rewarded and Feynman achieved the breakthrough that has shaped our understanding of quantum electrodynamics and other quantum field theories to this day.

3.1 Going to Cornell

While in Los Alamos, Feynman had already received offers for the period after the war – many people had recognized his extraordinary talent there. In 1943, for example, Robert Oppenheimer had already tried to convince the University of California in Berkeley near San Francisco to make Feynman an offer. But it wasn't until the summer of 1945 that Berkeley finally decided, and it was too late, as it turned out. Hans Bethe had been quicker off the mark. At the end of 1943, he arranged an offer from Cornell University, which is located in the small town of Ithaca, about 350 km northwest of New York City, in the heart of the state of New York (Fig. 3.1). In 1935, Cornell had offered Bethe a new home after his flight from Germany, and

J. Resag, *Feynman and His Physics*, Springer Biographies,
https://doi.org/10.1007/978-3-319-96836-0_3

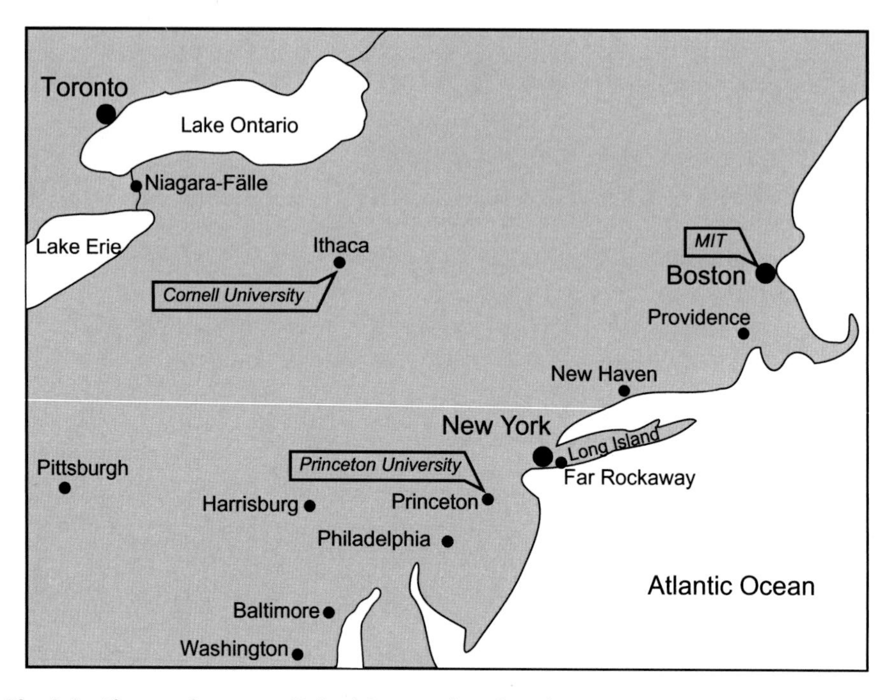

Fig. 3.1 The northeastern United States, showing the most important stages in the life of the young Richard Feynman. Near New York you can see the small town of Far Rockaway, where he was born and raised. At the top right is the MIT in Boston, where he first studied physics, before moving to Princeton to do his doctoral thesis. Following his involvement in the Manhattan project in Los Alamos, which is located in New Mexico in the southwest of the USA and cannot be seen here, he moved on to Cornell University in Ithaca

Bethe remained loyal to Cornell for the rest of his life. Feynman admired Bethe and was looking forward to working with him after the war, so he gladly accepted the offer. Feynman also very much liked Oppenheimer, but the latter was unable to compete.

In October 1945, Feynman left Los Alamos and took a train across the continent towards Ithaca. In the night train beyond Buffalo, he began to prepare his course on mathematical methods of physics, which he was to give shortly afterwards at Cornell University. Feynman was accustomed to such night-time working hours from his experience with the Manhattan project, where everybody had been working so hard and under great pressure to get everything finished at the last minute.

Arriving in Ithaca, Feynman realized that it was going to be very difficult to get a hotel room. So he left his suitcase in one of the hotels which agreed to look after it for him and started to wander around, looking for a room.

But there were no vacant hotel rooms anywhere, and he finally ended up exhausted at the university campus, where he saw an empty couch in the hallway of one of the buildings. Should he really sleep there? Now that he was a university professor, he wanted at least to be a little dignified – what if the students found him there sleeping on the couch? But it was late and he was tired, so he asked the janitor whether he could sleep on the couch – "Sure," he replied.

The next morning he hurried to the physics department – where was he to give his course? Was he late already? But his panic was completely unfounded! He was told that the courses would not start until a week later. He had only been called to Ithaca earlier so that he could look for a place to stay and get used to his new surroundings. What a difference to the hectic life at Los Alamos! "I was back to civilization, and I didn't know what it was", writes Feynman in *Surely You're Joking*.

He was pointed to the building where he had involuntarily stayed the previous night, and where he would be assigned a room. The clerk responsible for this thought he was a student and laughed: "Buddy, the housing situation in Ithaca is tough. In fact, it's so tough that, believe it or not, a professor had to sleep on a couch in this lobby last night!" Word had already got around, so Feynman replied: "Well, I'm that professor, and the professor doesn't want to do it again!"

When finally supplied with a room, Feynman invested a lot of time and effort over the following weeks to prepare his lecture on mathematical physics and another lecture on electricity and magnetism. They were certainly good lectures, because Feynman loved to teach and he was extremely demanding with regard to the quality of his lectures. Many years later he realized that he hadn't noticed how exhausting the preparation of his first courses must have been for him – he had lost any sense of how tired and depressed he was after the strains of the Manhattan project and the tragic death of his young wife.

Depressed and Burnt Out

When Feynman came to resume his research, he was surprised to discover that he was unable to accomplish anything. He had had many ideas before the war, but now he was tired and could no longer find any interest in it. He had the feeling that he would never be able to do any further scientific research. He was deeply depressed. Had he really reached the end of his scientific career? Was he already burnt out at the age of 27?

He was all the more ashamed in that he was still receiving lucrative offers from other universities – they just didn't realize that they wouldn't get anything for their money, or so he thought. This situation reached its climax when he had an offer from the Institute for Advanced Study in Princeton, where geniuses such as Einstein and von Neumann were currently working. The offer even included special conditions for Feynman, allowing him to give lectures at Princeton, because they knew that Feynman didn't believe in retreating into an ivory tower where he would have no obligations and would lose contact with students and experiments. Feynman was convinced that it was impossible to work productively without such stimulation from outside. So the offer had even been adapted to Feynman's own personal preferences.

My goodness, Feynman thought – this position was even better than Einstein's. And they had offered it to a young physicist who was unable to accomplish anything. "It was ideal; it was perfect; it was absurd!" writes Feynman, for whom it was out of the question to accept such an offer given the poor condition he was in at the time. He would never be able to satisfy anyone's expectations.

To top it all, Feynman had to cope with another serious blow coming just over a year after Arline's death. His father Melville, who had always been very close to him and who had been in poor health for some time, finally died of a stroke in October 1946. In the same month, Feynman wrote the letter to Arline in which he poured out his heart. She would understand him in his loneliness, even if he could no longer speak to her in person.

Fortunately, Richard had at least been able to make things up with his mother Lucille. After Arline's death, she had asked him to forgive her for not seeing things his way. She had been afraid for him and feared that Arline's fatal illness would be too much for him to bear. But he had borne it so well, she wrote in a letter, and tried to encourage him: "Now try to face life without her."

On the outside, Feynman managed to hide his depression quite well. When Hans Bethe heard about this later, he said: "Feynman depressed is just a little more cheerful than any other person when he is exuberant."[1] Neither he nor Feynman's office colleague Philip Morrison had noticed anything amiss.

[1]See, e.g., Silvan S. Schweber: *QED and the Men Who Made It.*

But one of his colleagues had indeed noticed Feynman's condition. This was Bob Wilson, who had talked Feynman into joining the Manhattan project and was now head of the nuclear physics laboratory at Cornell. Wilson called him into his office in the spring of 1947 and tried to reassure him. He said they were very satisfied with the way Feynman was teaching his classes: "You're doing a good job, and we're very satisfied!" Any other expectations they might have would just be a matter of luck, and any risk involved was entirely up to the university. He should not worry so much about it.

Feynman felt relieved. It didn't make any sense to focus constantly on the expectations of others. Perhaps he also remembered what John von Neumann had told him in Los Alamos: "You don't have to be responsible for the world that you're in" – and neither therefore did he need to hold himself responsible for other people's expectations. He was what he was, and if they expected him to be good and offered him money, that was not his problem.

Physics Becomes Fun Again

And then Feynman began to remember how much fun physics had always been for him. He had simply treated it as a game, giving little thought about whether it was important for the development of science. And now he wanted that fun back! If he was burnt out anyway and couldn't achieve anything important in the way of research, he might just as well go on playing with physics as he had always done in the past. He had a nice position at the university and was allowed to teach students, something he enjoyed very much – so what could go wrong with that?

That was exactly the attitude he needed to recover his creative power. Just for the fun of it, he calculated how a plate thrown into the air would rotate and wobble, because that's exactly what he had observed accidentally in the cafeteria. He had noticed that the plate was rotating twice as fast as it was wobbling, and he would not drop the subject until his calculations had confirmed this observation.

When Bethe asked him what importance that might have, Feynman just said: "Hah! There's no importance whatsoever. I'm just doing it for the fun of it!" He didn't care if other people thought such games were important for physics or not. He simply wanted to enjoy the subject again. But as often happens, one thing led to another. He thought about wobbling plates, then about rotating electrons and their relativistic movement, then the Dirac equation, and finally quantum electrodynamics. And before he realized it, he was playing around with the old problems that had occupied him before his time at Los Alamos. It was like uncorking a bottle – everything flowed out effortlessly, he remembers later. The young Feynman, brimming with ideas, was back again!

Not all his ideas worked right away. In his Nobel Prize speech, Feynman later said that he spent almost as much effort on the things that didn't work, as on the things that did work. This is of course perfectly normal at the forefront of research. In most cases, there is hardly anything that actually works, and good ideas that really carry can often be put down to a stroke of luck.

We already mentioned one of the failures in the last chapter. The theory of electrodynamics without self-interaction that Feynman and Wheeler had established could not deliver a consistent quantum theory. Feynman had already discovered this on his many journeys during the war, on those occasions when he had had some time for his research.

He also encountered some problems with the relativistic description of the electron spin in his path integral formalism. In one space dimension he succeeded in deriving the corresponding Dirac equation. But he failed in three dimensions. This was unsatisfactory, because the non-relativistic Schrödinger equation could be derived without any problems from the path integrals.

Feynman tried to organize his thoughts by compiling the results of his doctoral thesis – his path integral method – for publication. He hated this kind of work, for it forced him to give up his rather intuitive approach and arrange his thoughts according to a strict logical structure. Feynman loved to use ingenious ideas to solve problems, with the end sometimes justifying the means. However, he was not particularly interested in the rather formal approach required for publication in a physical journal. But in the end he succeeded and the paper appeared in April 1948 under the title *Space-Time Approach to Non-Relativistic Quantum Mechanics*.

Feynman became more and more accustomed to thinking in "path integral pictures", because of his intense involvement with this topic. A particle took every possible path in space and time, and every path contributed a wave amplitude to the overall amplitude. Paths and amplitudes would soon become his universal tools.

The problems with the Dirac equation showed that it was not so easy to incorporate relativity theory into quantum theories. However, this was absolutely necessary for quantum electrodynamics. So what specific difficulties did this entail?

One of the most important findings of relativity theory is the equivalence of mass and energy. Mass is indistinguishable from stored energy. For example, 99% of the mass of a nucleon (proton or neutron) is generated by the strong field that holds the three quarks together, while the quark masses contribute only about 1% to the nucleon mass. So today we can say that almost the entire mass of atoms comes from the energy of this strong field.

In physical processes, the total energy of all particles must be conserved. Energy can therefore neither be created nor destroyed. With mass the situation is quite different! It does not have to be conserved, even if that corresponds to normal experience. For example, if a uranium-235 nucleus is split by a neutron into a barium-142 and a krypton-92 nucleus plus two neutrons, about 0.08% of the mass is lost. The energy stored in this mass is released by such a nuclear fission reaction and appears as kinetic energy of the fission products. Mass is apparently a very effective energy storage device, because even this relatively small mass loss produces the enormous explosive force of an atomic bomb. Einstein's formula $E = m \cdot c^2$ reflects this, because the speed of light c is so great, and it is squared in the formula, ensuring that even a small mass m will correspond to a very high energy E.

Mass can thus be converted into other forms of energy. The reverse process is also possible, as large particle accelerators show every day. Here, for example, protons are shot at each other with extremely high momentum, and the available energy is used to form a host of new particles which usually decay again very quickly. The equivalence of mass and energy ensures that the number of particles can change during a process. Particles can be created and then disappear again.

As a consequence, in Feynman's path integral method, it makes no sense to confine oneself to the possible paths of a single particle, if we are trying to incorporate the theory of relativity. We must also include processes in which new particle paths are created and others disappear.

The Problem of Negative Energies

There was another problem that had to be solved: both the Dirac equation and the Klein–Gordon equation – the two most important relativistic quantum equations – predict negative energies, i.e., energies that are less than zero. This is because these equations both imply the relativistic relationship $E^2 = (m \cdot c^2)^2 + (p \cdot c)^2$ between particle energy E and particle momentum p (Fig. 3.2, we omit potential energies to simplify).

Since the energy is squared, both positive and negative energy values can satisfy this equation (the sign disappearing when we take the square). Normally, we might simply insist that the energy E be zero or positive, because we don't know what a negative particle energy is supposed to mean.

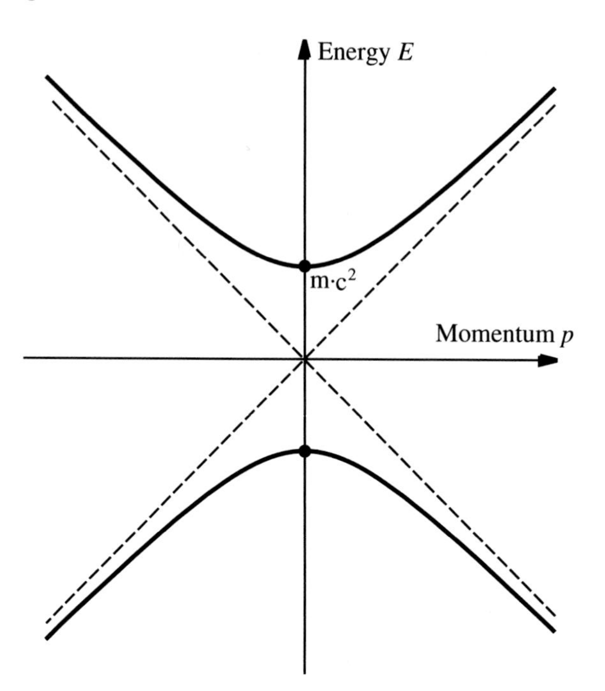

Fig. 3.2 The upper solid line shows the relativistic relationship between the energy E and the momentum p of a particle. For a particle at rest, the momentum is zero and the energy is equivalent to the mass m of the particle according to Einstein's formula $E = m \cdot c^2$. The lower curve corresponds to energy values with opposite signs, as required by relativistic quantum theory

However, the quantum equations also make mathematical sense for negative energies, so quantum waves exist as solutions for both negative and positive energies.

Well, why don't we just leave out the waves with negative energies? But that just isn't possible! If, for example, a quantum wave with positive energies enters a force field, then quantum waves with negative energies are inevitably generated. The same thing happens if we try to create a spatially bounded wave packet – the smaller it gets, the more significant the negative energies become. Negative energies cannot be ignored.

Can these quantum waves with negative energy correspond to real particles, as they do for positive energies? This is hard to imagine, because if we could produce particles with negative energy, our world would be unstable. Unlimited numbers of particle pairs could arise out of nothing, one particle having a positive energy and the other particle a negative energy of the same amount. In sum, the total energy of the particle pair would be zero, so no energy would be needed to create it. Empty space, also called the vacuum, could be filled with any number of such particle pairs.

This is clearly not what we observe in the real world. Is there any way to save the situation?

The Dirac Sea

When Paul Dirac formulated the Dirac equation in 1928 and was confronted with the problem just described, he came up with an ingenious idea just two years later: what if one did not define empty space as the state that contains no particles, but as the state with the lowest total energy? In order to achieve this state, all quantum states of negative energy would have to be filled with particles, so that there is no longer an unoccupied state of negative energy. The vacuum would be filled with a sea of particles of negative energy – the so-called Dirac sea.

Since the Dirac equation describes particles with half-integer spin (quantum angular momentum) – for example electrons or quarks – there would simply be no room for further particles in this sea. This is because, for such particles, each quantum state can only be occupied by a single particle due to the Pauli principle, which some readers may already know and to which we shall return in the next chapter (see Sect. 4.2). So, if all negative energy states are already occupied in the vacuum, only positive energy states remain available for further particles. No new particles with negative energy can be generated from the vacuum.

To make all this physically meaningful, we have to assume that the filled Dirac sea appears to us exactly like empty space without any particles. We may imagine, for example, that the sea fills space evenly, so that we do not perceive it – just as we cannot see the water itself when we are under water.

However, the filled Dirac sea would have an infinitely large negative energy, because there is an infinite number of quantum states in it. This means that we would have to reset the zero energy and measure all energies relative to the filled Dirac sea, whence the latter would have zero energy by decree. In quantum electrodynamics this would be allowed, because only energy differences are relevant there, so any energy can be used as a reference value. We will soon encounter this idea again when we come to consider renormalization.

This approach becomes problematic if we include gravity. Since energy and mass are equivalent according to Einstein, each energy generates a gravitational field, so we can no longer shift the zero point of the energy in an arbitrary manner. Hence, the idea of the Dirac sea thus becomes questionable when gravity comes into play. Fortunately, gravity is so weak in the realm of atoms and particles that we can safely ignore it there. What happens when gravity and quantum theory become important at the same time we will get to know in Sect. 5.4.

Another situation where the idea of a filled sea of negative energy particles no longer works occurs when we look at particles with integer spin (e.g., photons or mesons), as described for example by the Klein–Gordon equation. Here too there are negative energies, but the Pauli principle does not apply to these particles, so the negative energy states cannot be completely filled up – there is still room for further particles in every quantum state. For this reason, the Klein–Gordon equation was considered questionable for many years and only the Dirac equation was thought to be correct.

So the idea of the Dirac sea wasn't fully satisfying, but in 1930 it was the best available to deal with these negative energies – at least in the case of particles with half-integer spin. Since all the building blocks of the atom – electrons, protons, and neutrons – are just such particles, there didn't seem too much cause for concern.

Antiparticles and Virtual Particles

In addition, the Dirac sea opened up a fascinating possibility. Imagine that one would supply one of the particles in the sea – let's say an electron – with enough energy to lift it from its state of negative energy to a state of positive energy. The result would be an ordinary electron with positive energy and a missing electron with negative energy – a hole in the Dirac sea (Fig. 3.3). The Dirac sea would then have a negative charge and a negative particle energy less than before, i.e., the hole would look like a particle with positive charge and positive energy relative to the intact Dirac sea. This particle would have exactly the same properties as an electron, only with the opposite charge. Such a particle would be referred to as the *antiparticle* of the electron, or because of its positive charge, simply the *positron*.

So taken together, lifting an electron out of the Dirac sea would look as if the energy supplied had materialized in the form of an electron–positron pair. When Dirac became aware of this possibility, it raised a doubt – positrons had never been observed before. So he speculated that the mass of the positive hole in the sea might increase through its interaction with the rest of the sea, whence it might be interpreted as a proton. However, oth-

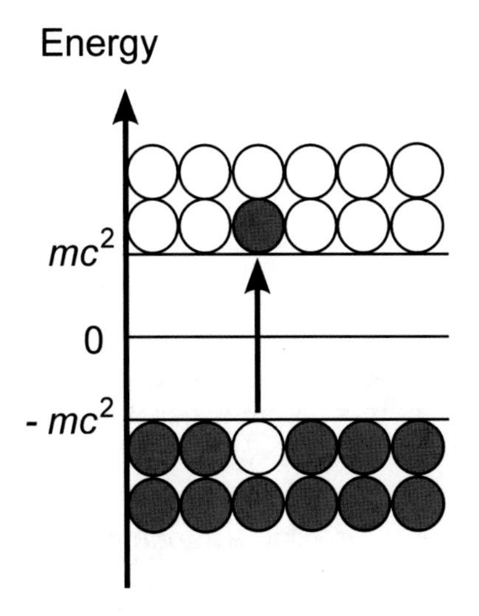

Fig. 3.3 When an electron is lifted from its negative energy state, a hole is created in the Dirac sea. Such a hole is interpreted as a positive anti-electron (positron)

ers, including Robert Oppenheimer, objected that this idea could not work because the hydrogen atom, being made of a proton and an electron, would not then be stable. The electron and proton could destroy each other by the electron dropping into the proton hole in the Dirac sea. This was also clear to Dirac, and he finally conceded to the existence of positrons.

If only Dirac had dared to predict the positron from the very beginning! A short time later – in 1932 – this very particle was actually discovered by Carl David Anderson in cosmic radiation. "My equation was smarter than I was", Dirac commented when asked about his initial hesitation.

There is a third complication when it comes to combining relativity and quantum theory. It has to do with the fact that the shorter the lifetime of a quantum state, the less precisely its energy is determined. For very short periods of time, this energy uncertainty can become so great that it can induce the materialization of new particles. These so-called virtual particles borrow the necessary energy needed for their existence by exploiting the time–energy uncertainty, but they must then return this energy immediately afterwards and thus end their ultra-short lives.

When we consider the idea of the Dirac sea, this means that over very short periods of time we have stormy waters before us rather than a calm lake. Electrons are constantly being lifted from their negative energy states for a moment and then immediately falling back again. Only over longer periods of time do these effects average out in such a way that the sea appears calm.

Electric charges sense this fluctuation of electrons in the sea. For example, a negative charge would repel the raised negative electrons and attract the positively charged holes. Empty space is no longer only the stage for physical processes; its quantum properties make it a physical actor that influences these processes.

This makes the whole matter very complicated. There is a sea filled with negative energy states, energy can transform into particles and vice versa, and for short periods of time, the energy–time uncertainty means that countless virtual particle–hole pairs have to be taken into account. (Incidentally, there is another, even better known quantum mechanical uncertainty, which we will learn more about in Sect. 4.2.)

At that time, calculations in quantum electrodynamics were mostly done by analogy with the non-relativistic Schrödinger equation, i.e., physicists dealt with wave functions and their time evolution. In addition, there was a relatively abstract formalism with mathematical operators that could generate and destroy particle states.

Feynman didn't like this formalism very much. How could an operator generate an electron? Wouldn't that contradict charge conservation? This was not the case, of course, because for every negative charge that was generated, a positive charge was also created. Nevertheless, Feynman initially refused to learn this very practical and widespread, albeit somewhat abstract, way of doing the calculations: "I blocked my mind from learning a very practical scheme of calculation", he says in his Nobel Prize speech. This reminds us very much of the younger Feynman, who in his studies refused to apply the elegant Lagrangian method of classical mechanics, preferring to calculate everything his own way.

But it was precisely this stubbornness that enabled him to avoid some of the problems others were struggling with. Using what was then the most common calculation method, it was very difficult to keep track of the whole set of equations one was dealing with. Furthermore, it was not clear whether the theory behaved correctly when one switched from a static to a steadily moving observer. One of the basic rules of the special theory of relativity is that the laws of physics must look the same under such a change, since all such observers are on an equal footing. So there is no observer who is really at rest in an absolute sense – or put another way, all non-accelerating observers can declare themselves as being at rest. However, it was not easy to see how this basic rule could be respected in the equations, as they seemed to change dramatically – at least at first sight – when switching from one observer to another.

Infinities Everywhere

To top it all, infinities turned up all over the place in the formulation of quantum electrodynamics, for example in the self-energy of the electron, which Wheeler and Feynman had originally wanted to eliminate with their approach.

In short, the whole theory seemed to be inconsistent, and most physicists were beginning to get quite frustrated. Could the whole structure still be saved somehow, or would a completely new theory be required?

This was where Richard Feynman came in. He and a few others would eventually succeed in putting the theory on a firm foundation. It was only necessary to look at it from a different angle, with a deeper understanding of the physics behind the thickets of formulas. It was not the time development of wave functions that would provide the means of choice. It was rather the complete picture of emerging and decaying particles in space and time that eventually led to the desired goal.

Feynman also required a certain amount of time to find the right way to do this. He played around with a variety of approaches and over the course of time received more and more practice in formulating quantum electrodynamics in somewhat different ways. He knew many of the standard methods available at that time, but he continued to experiment with his path integrals and found completely new perspectives. These were the two most productive years of Feynman's scientific career.

3.2 The Masterpiece: Feynman Diagrams and Antiparticles

Paths and associated amplitudes were the basic tools in Feynman's world of thought. But how could quantum electrodynamics (QED) be expressed in this language? How the demands of special relativity theory be included, and what about negative energies and virtual particles?

It happens that there is a very useful method for tackling these questions in a thoroughly systematic way. This method is known as perturbation theory. It is also used in non-relativistic quantum theory and applies equally well to QED. The idea is that interactions are taken into account step by step.

Disturbed Quantum Waves: Non-relativistic Perturbation Theory

Here is a typical example from non-relativistic quantum mechanics: a scattering process for which Feynman's way of thinking is particularly suitable. It starts with a beam of free particles – let's say electrons – which we can describe by means of a plane quantum wave moving freely through space. This well-defined waveform is "disturbed" when it meets the electric field of an atomic nucleus, for example. The plane wave fronts are slightly bent and scattered waves are created, leaving the region of the atomic nucleus on all sides. Far away from the atomic nucleus, these scattered, almost spherical wave fronts move away undisturbed once again – they correspond to electrons, which have been thrown off their previously straight course by the field of the atomic nucleus (Fig. 3.4).

It is very easy to describe the undisturbed motion of a wave mathematically. However, this becomes difficult in the interaction region where the electric field acts on and disturbs the wave. This is exactly the region where perturbation theory is applied. It describes how the motion of the wave is influenced by the field, which creates new elementary waves everywhere as the wave passes through it. These new elementary waves then move on undisturbed once again. The stronger the field and the passing wave in a given place, the stronger will be the new elementary wave generated there.

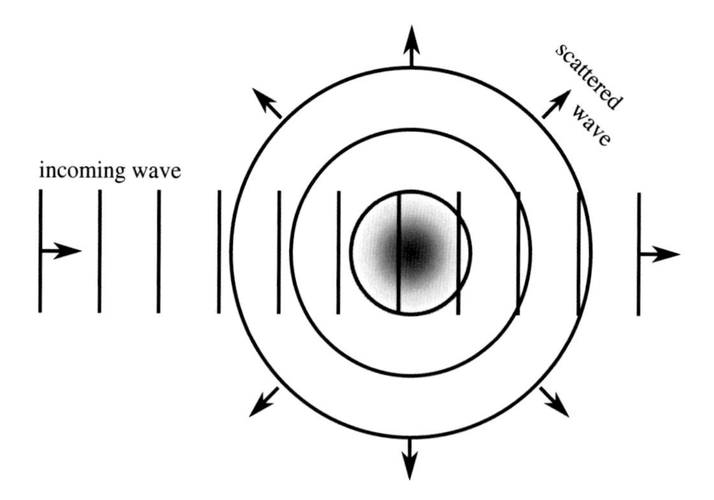

Fig. 3.4 Scattering of an incoming plane electron wave in the electric field of an atomic nucleus (grey region)

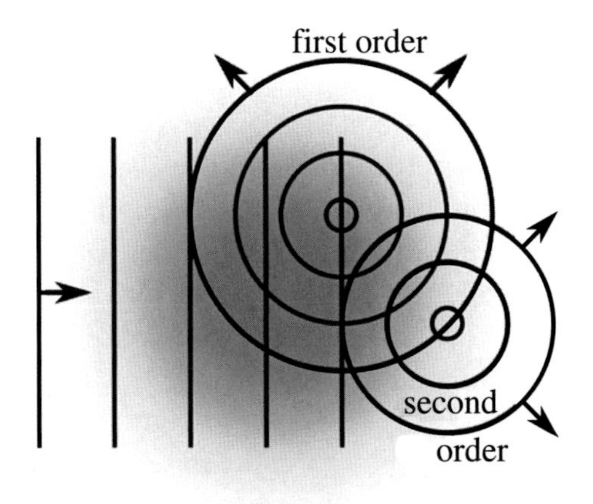

Fig. 3.5 Elementary waves of first and second order in perturbation theory

All these new elementary waves then interfere with each other and with the original unperturbed wave to form the total wave disturbed by the field.

In the simplest case, an elementary wave is thus generated at any point in the field and at any time the wave passes through it, and then propagates undisturbed further on. This is called the first order of perturbation theory. If the field is not too strong, a good approximation is thereby obtained for the overall result.

If we want to know it more precisely, or if the field of the atomic nucleus is stronger, we have to take into account the fact that the newly generated elementary waves are also influenced by the field. Not only the original wave, but also the new elementary waves create further elementary waves, which in turn spread undisturbed.

In second order perturbation theory, the total wave thus consists of three parts (Fig. 3.5): the undisturbed original wave (zeroth order), the elementary waves of first order recently generated by it as it interacts with the field, and the elementary waves of second order generated by these waves when they in turn interact with the field.

Of course this game can be continued forever, because the secondary elementary waves can also create new elementary waves, and so on. The more orders are taken into account, the smaller the higher order corrections should become and the more accurate the result. This works well with electric fields, because the forces acting there are usually weak enough to ensure that the elementary waves of higher orders really do become weaker.

In order to get an overview of all these processes, perturbation theory can be illustrated by space-time diagrams. To simplify things we restrict ourselves to one space dimension, which we label as the x-axis. The vertical axis is then used to represent time t.

Imagine that at a time t an elementary wave is created at a location x and propagates from there undisturbed – our undisturbed electron beam consists of precisely such elementary waves. At a later time t' this wave has certain values – amplitudes – at other locations. Let us pick one of these locations and call it x'. In the diagram we draw a straight line from the starting point (x, t) to the end point (x', t'), because this is the path covered by the corresponding part of the elementary wave in space and time between the two points. This line represents the zeroth order of perturbation theory.

For the first order we need a line from the starting point (x, t) to an intermediate point (x_1, t_1) where a new elementary wave is created, and then another line from this intermediate point to the end point (x', t') which represents the undisturbed progress of this new elementary wave. In second order, we would have two such intermediate points, and so on (Fig. 3.6).

This is already a good start toward obtaining an overview. It is tempting to imagine the lines in the diagrams as real particle paths, but this is deceptive, because they represent only the propagation of elementary waves. However, one can reasonably think of them as possible particle paths, in the sense of Feynman's path integrals. There, a particle – or rather the associated quantum wave – takes any possible path between two points, each of these paths being associated with a quantum amplitude which represents its contribution to the quantum wave at the end point. All these amplitudes are then summed to obtain the total quantum wave at the end point, and it is in this sense that each path contributes with its amplitude.

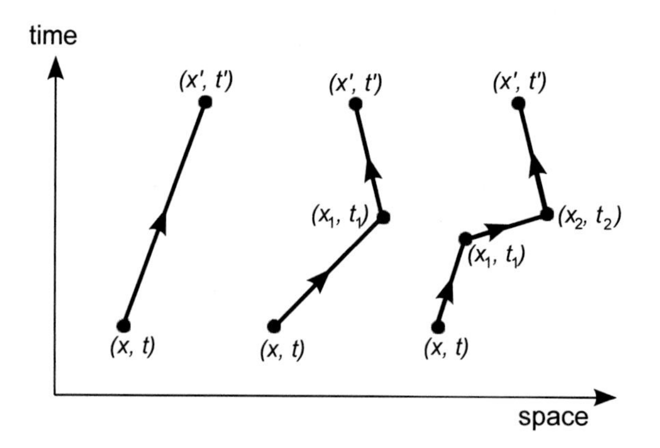

Fig. 3.6 Diagrams of non-relativistic perturbation theory up to the second order

This is exactly how things work here, too! Each diagram represents possible paths between the start and end points: the path without a kink, all paths with one kink, all paths with two kinks and so on, where the kinks can be located anywhere in the electric field of the atomic nucleus. Each of these paths has a quantum amplitude, and we then sum over all paths, including all possible intermediate points, to calculate the quantum wave at the end point.

The advantage with perturbation theory is that the corresponding path amplitudes can actually be calculated. The straight sections of the paths belong to the undisturbed propagation of an elementary wave, for which there is a relatively simple formula. These amplitudes must then be multiplied together to obtain the amplitude of the entire path.

The kinks – i.e., the intermediate points – must also be taken into account when evaluating the amplitude of a path. The next elementary wave is generated at these points by the electric field of the atomic nucleus, and the stronger this field is there, the greater the corresponding amplitude of this wave.

Now the strength of the electric field depends on the electric charge Q of the atomic nucleus. Each kink thus contributes a charge factor Q to the total amplitude of the path. Thus, the amplitudes of the paths with one kink contain one factor of Q, those with two kinks contain two factors of Q, i.e., Q^2, and so on. This also makes it clear that the charge Q must not be too large for the whole process to work. For small charges Q, the amplitudes of the paths become smaller and smaller, the more kinks, i.e., Q-factors, they contain, so that such calculations can be limited to paths with only a few kinks.

Perturbation theory is therefore an approximation method. It can be used to approximate a complicated situation that cannot be calculated in an exact way. To do this, we take advantage of the fact that the problem contains a small parameter – here the charge Q of the atomic nucleus. The procedure then produces a term with Q plus a term with Q^2 plus a term with Q^3 and so on. Each of these terms can be calculated. The only thing is that we must stop somewhere, because we can't handle an infinite number of terms. However, if the charge Q is sufficiently small, the higher the power of Q in a given term, the less significant the term will be. So we can indeed stop somewhere and still get good accuracy for the overall result.

Perturbation Theory in QED: Feynman Diagrams

So much for non-relativistic perturbation theory. What changes if we switch to quantum electrodynamics (QED), where the theory of special relativity

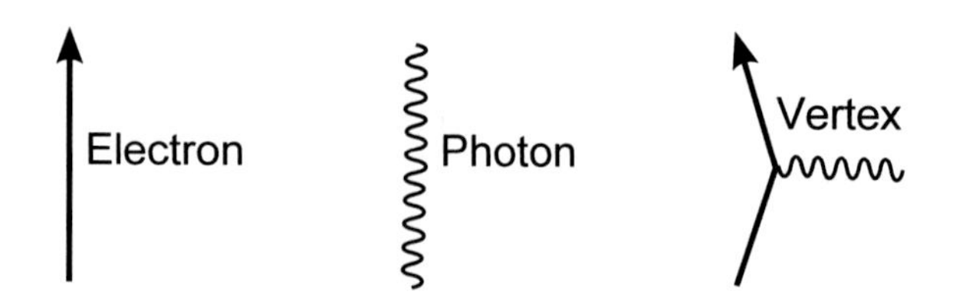

Fig. 3.7 Components of a Feynman diagram in QED

comes into play? Can perturbation theory still be successfully applied in QED? Feynman devoted himself to this question and developed the corresponding space-time diagrams, which have since become widely known as Feynman diagrams.

First of all, Feynman had to replace the classical electromagnetic field with a quantum description, introducing a new type of particle: electromagnetic field quanta known as photons. The kinks in the paths of the electrons then no longer result from the influence of a classical electric field, but from the influence of a photon that is emitted or absorbed at the kink. Just as in classical theory where electromagnetic waves are emitted or absorbed by charged particles, this happens with photons in QED.

In his diagrams, Feynman represented the elementary wave of a photon by a wavy line in order to distinguish it better from the solid lines of the electrons. A kink – henceforth called a *vertex* – is now a point where a wavy photon line branches off (Fig. 3.7). A vertex thus marks a point where the elementary wave of a photon is newly created or absorbed by an electron.

In the corresponding formulas, this means that the charge e of the electron takes the place of the nuclear charge Q, because this charge determines how strongly the electron interacts with a photon. However, we do not specify the charge in any artificial unit like the coulomb. Nature gives us the possibility to construct a pure number from the (squared) charge using the speed of light c and the Planck constant $\hbar$. For historical reasons, this is called the fine-structure constant and denoted by the letter α:

$$\alpha = \frac{e^2}{4\pi\varepsilon_0 \hbar c} \approx \frac{1}{137}.$$

Here ε_0 is the electric constant, which is needed as a conversion factor when specifying e in coulomb.

The value of α is independent of the units of measurement, so it is also called the (squared) electron charge in natural units. Fortunately, with its value of about 1/137, this number is relatively small, so the charge of the electron in natural units, i.e., the square root of α, which is about 0.085, is also quite small compared to unity.

This is exactly what we need for perturbation theory! Each vertex in a diagram results in a factor of 0.085, so the more vertices it contains, the less the diagram will contribute. In most situations, we can thus limit ourselves to diagrams with only a few vertices. This is a great advantage, because the calculations become extremely complex when more vertices come into play, the reason being that complicated integrals arise with each additional vertex.

Next, Feynman had to find out what the elementary waves of the electrons and photons look like, because they determine the amplitudes that correspond to the straight lines in the diagrams. In the case of photons, this seemed relatively simple at first glance: elementary photon waves should propagate at the speed of light, because photons are quanta of light.

However, Feynman's calculations showed that this point of view was not quite correct: a small part of the elementary wave is somewhat slower than the speed of light and another small part is slightly faster, in such a way that the amplitude of the elementary wave becomes smaller the greater the deviation from the speed of light. In his book *QED: The Strange Theory of Light and Matter*, Feynman writes: "You found out earlier that light doesn't go only in straight lines; now, you find out that it doesn't go only at the speed of light!"

At long distances, the amplitudes for being faster or slower than the speed of light are very small compared to the contribution corresponding to the speed of light itself. In fact, they cancel out when light travels over long distances, as Feynman goes on to explain. So no one would ever notice a deviation from the speed of light in our everyday environment, and no one would ever be able to send signals faster than the speed of light – the theory of relativity is thus saved. In the subatomic world, however, we must take these possible deviations from the speed of light into account.

What does this mean for the Feynman diagrams? To answer this question, we first scale the x- and y-axes so that a photon with light speed corresponds to a diagonal line in the diagram. If a photon line is steeper than the diagonal, it corresponds to a wave component that propagates more slowly than the speed of light, and if it is less steep, it represents a wave component moving faster than the speed of light.

For photon lines that start in the diagram at one vertex and end at another, all angles of inclination are now allowed, i.e., all possible photon paths must be considered, not only diagonal ones (Fig. 3.8). So we shouldn't be surprised if we even find horizontal photon lines in the diagrams.

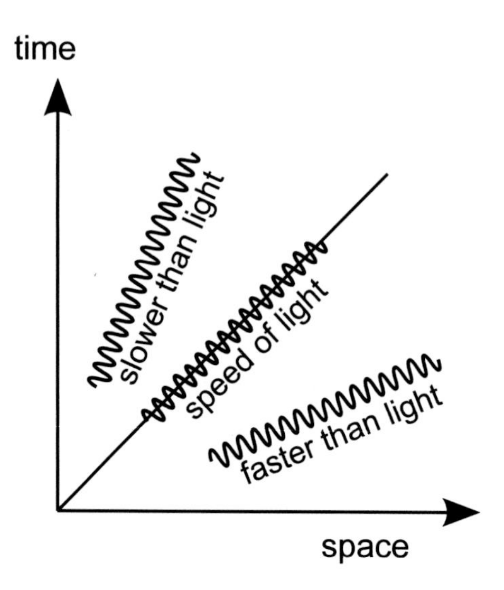

Fig. 3.8 Photon lines with different velocities in the Feynman diagram

If a photon appears only as an internal line in a diagram, it does not exist as a real photon, because it does not leave the diagram. It represents an electromagnetic interaction at the quantum level, but is not itself perceptible as a particle, precisely because for this it would have to leave the diagram. Such photons are said to be *virtual*, because – like the virtual particle–antiparticle pairs discussed earlier – they live only for a short moment and do not exist as real measurable particles. Therefore, virtual photons can travel faster or more slowly than the speed of light, unlike the long-lived photons that can be measured experimentally.

For virtual photons, it does not matter in which direction we imagine their propagation, because the formula for the corresponding amplitude contains both vertices in a completely interchangeable way. Such a photon is thus said to be *exchanged* between the two electrons.

The situation is different with electrons, which is why electron lines in the diagrams are marked with a small arrow pointing more or less upwards in the direction of time. In addition, individual electron lines cannot start or end at a vertex, because this would violate the law of conservation of electric charge (we discuss the generation of electron–positron pairs below). In other words, electron lines always run right through a diagram and are only kinked at the vertices.

If we once again draw the diagrams in such a way that the diagonal corresponds to the speed of light, we normally expect electron lines to be steeper than the diagonal, because according to Einstein's theory of relativity, no

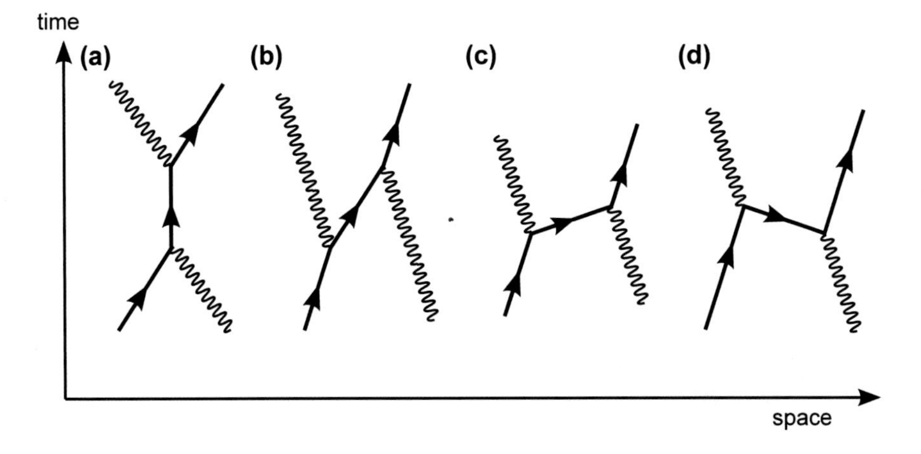

Fig. 3.9 Four diagrams for the scattering of light by electrons

real particle can move faster than light. But beware! We have just seen that photons can move faster than light over very short distances and the same applies to electrons! Once again, the corresponding elementary waves have a small component that exceeds the speed of light. This is especially important between two vertices – here an electron line in the diagram can also have a very shallow slope, i.e., move much faster than the speed of light.

Figure 3.9 shows a process in which light is scattered by electrons. There are various possibilities. In the simplest case, a photon is absorbed by an electron and later emitted (a). However, the electron could also first emit a photon and then absorb another one (b). Both are quantum mechanical possibilities. In diagram (c), the electron line between the vertices is less steep than the diagonal, so the electron is moving faster than the speed of light. The vertices in the diagram are located in such a way that, in the short time between emission and absorption of the photons, the spatial distance between the two locations can only be crossed by an electron with superluminal speed. This is not a problem, because the electron does not appear as a real measurable particle between the two vertices – the distance is much too small. Between the two vertices, the electron can thus be regarded as a virtual particle.

For those vertices whose spatial separation can only be crossed with superluminal speed, the special theory of relativity makes an astonishing statement: their time sequence can be reversed when viewed from the perspective of an observer moving sufficiently fast in the right direction. According to Einstein, since all observers who move uniformly relative to each other are completely equivalent, there is no absolutely binding chronological order for the two vertices. One observer would say that a photon is first emitted and then another photon is absorbed. Another observer would see it the other way round as in diagram (d). There is even an observer who considers both events to occur at the same time. This is why we speak of the "relativity of simultaneity".

However, the observer, who considers diagram (d) to be correct, sees something strange: on its way between the two vertices, the electron seems to move backwards in time. Only, what could that mean? Is such a thing allowed?

Back to the Future

We have seen before that it may well make sense to allow the time direction to be reversed. This was exactly what Feynman and Wheeler had done when they thought about advanced electromagnetic fields and waves running backwards in time. The basic laws of physics are almost completely symmetric with regard to the direction of time, so they do not distinguish between the future and the past (except for a rare kind of exception which affects certain particle decays and which we can safely ignore here) – a reversal of the direction of time seems therefore quite legitimate.

On the other hand, in our macroscopic world there is a clear difference between the future and the past. Time is like an arrow pointing in only one direction. None of us can simply travel back into the past against the arrow of time, although the basic laws of nature do not distinguish between the future and the past. So how does all this fit together?

It seems most likely that the preferred time direction does not arise directly from the fundamental laws of nature, but comes about rather for purely statistical reasons. The difference only occurs when we look at systems with a large number of particles, i.e., macroscopic systems, such as our own bodies.

We have already encountered this argument in Chap. 2, in the discussion between Einstein and Ritz: the macroscopic states of the past are simply much less likely than the macroscopic states to which our world evolves on its own. A cup of hot water cools down until it reaches room temperature – it will not get hotter than the room all on its own. It is much more likely that the water molecules of the hot water will release energy to the usually slower molecules of the air by random collisions with them, whereupon the energy will subsequently dissipate still further. But there will also be some very fast molecules in the room that can transfer energy to the water molecules, and sometimes they will actually do so. By an unbelievable coincidence, the motion of these fast molecules in the room could occur in such a way that they deliver a great deal of energy to the water molecules, whence the water would end up becoming much warmer than the room. However, this is so extremely unlikely that we will probably never see it. For all practical purposes, the cup always cools down to room temperature and will not suddenly get warmer than the room through chance collisions.

So the difference between the past and the future, according to today's knowledge, can be explained solely on the basis of such probability arguments. In the quantum physics of individual electrons and photons, however, the direction of time plays no fundamental role.

In our diagrams we can therefore allow electrons to go backwards in time. But what does this mean for a macroscopic observer whose time direction is always from the past to the future? What would such an observer see?

Why There Are Antiparticles

Conservation of charge provides a clue. If we follow the electron line in Fig. 3.9d, it just zigzags through the diagram. Nowhere is a new electron either created or destroyed. However, the macroscopic observer sees the chronological sequence strictly in the usual time direction, i.e., in the diagram from bottom to top. At first, the observer sees how an electron from the left and a photon from the right move towards each other. But before the collision, the electrically neutral photon on the right suddenly disappears and creates two particle lines that continue to run upwards. The line to the right with the arrow pointing up is another negatively charged electron. Since charge is conserved, the flat line in the middle with the arrow pointing downwards – the electron running back in time – must be a positively charged particle, if we look at it in the normal time direction. Since it is otherwise in every way similar to an electron, we should clearly assume that it is an anti-electron – a positron. Electrons moving backwards in time thus look like positrons when we look at them in the normal time direction. At the next vertex on the upper left, the positron and the original electron coming from the left destroy each other and a photon is created.

Let us summarize our thoughts once again. Like virtual photons, virtual electrons can also move faster than light, and this can also be justified mathematically if we take a closer look at the propagation of the corresponding elementary waves in quantum theory. In the theory of relativity, however, the chronological sequence of events depends on the observer, if these events are so close together in time that their spatial separation can no longer be travelled even at the speed of light. In the case of superluminal particles, one observer sees an object moving forward in time, while another observer sees an object moving backwards in time. In the normal time direction, we interpret a particle running backwards in time as an antiparticle. "One man's virtual particle is another man's virtual antiparticle", writes Feynman in his Dirac Memorial Lecture on the topic *The Reason for Antiparticles*, where he presents precisely the above argument. As a result, antiparticles are inevitable when we combine quantum theory and relativity theory. By the way, photons are identical with their antiparticles, because there are no properties that could distinguish photons from antiphotons – this is why the corresponding lines in the diagrams do not get an arrow.

For Feynman, the idea of looking at backward-running electrons was not a completely new one, as he and John Wheeler had already looked at backward-running (advanced) electromagnetic waves when he was at Princeton. Wheeler had gone even further at that time. As Feynman tells us in his Nobel Prize speech, he received a call from Wheeler in the spring of 1940: "Feynman, I know why all electrons have the same charge and the same mass. Because they are all the same electron!"

This idea was typical of Wheeler – as ingenious as it was crazy: according to Wheeler, there is only a single electron in the universe, moving backwards and forwards through time as it goes its way. Its classical path in space and time would be like a gigantic zigzag (Fig. 3.10). An observer would thus see many electrons moving into the future, as well as many electrons moving into the past. Such an observer would not see that it was always the same electron.

Of course, Wheeler also wondered what these electrons running backward in time would look like to the observer who looked at them in the usual time direction. Wheeler's consideration was based solely on classical electrodynamics in its relativistic formulation, without any quantum mechanics. For this purpose, he looked at the relativistic equation of motion of an electron in an external electromagnetic field. In this equation there was a special time parameter called *proper time*. This is the time that would be displayed on a clock that the electron carried along with it. This proper time does not have to be synchronous with the time of an external observer who looks at the electron's path and depicts it in the corresponding space-time diagram. On the contrary, as soon as the electron moves relative to this observer, its proper time runs more slowly than the observer's time, a phenomenon known as *time dilation*. The magnitude of the slow-down is given by the Lorentz factor γ which we already encountered in Chap. 1. In the

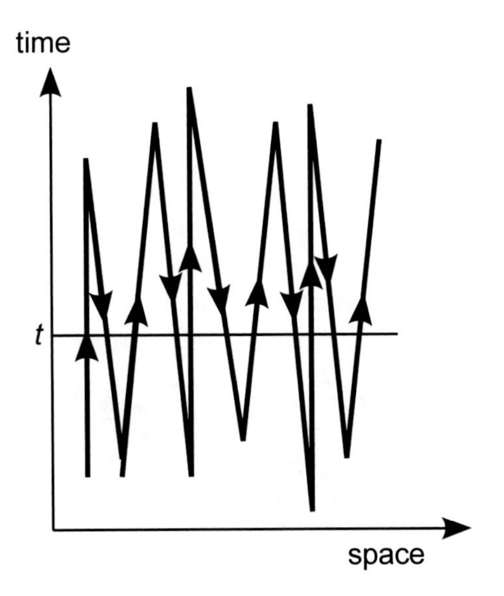

Fig. 3.10 Wheeler's one-electron universe. The arrows on the electron line indicate the direction of increasing proper time on a hypothetical clock carried by the electron. An observer at time *t* (horizontal line) interprets each line coming from below as an electron and each line coming from above as a positron

case of a particle that moves as fast as light, this slow-down becomes infinite, and no time at all passes for such a particle. In other words, a free photon never grows any older.

Mathematically, in this equation of motion the proper time can be run backwards relative to the observer's time. In our space-time diagram, the proper time of the electron then grows downwards, and this is indicated by arrows pointing downwards. The corresponding equation of motion, on the other hand, looks exactly like the equation of an electron whose proper time grows normally parallel to the external time – provided that one changes the sign of the charge of the electron so that it corresponds to a positron. An electron that travels backwards in time is therefore mathematically equivalent to a positron that moves forward in time. This is exactly the same result as we got above.

Wheeler's idea was fascinating, but Feynman still had doubts. If there were only one electron line that meandered through space and time, wouldn't there have to be as many positrons as electrons? Well, maybe the positrons are hidden in the protons of the atomic nuclei or something, Wheeler speculated. Feynman was not convinced, but the fact that positrons can also be understood as electrons moving backwards in time remained in his memory. "That, I stole!" confessed Feynman in his Nobel Prize lecture.

Matter and Antimatter – The Tiny Difference

In Feynman diagrams, electrons and positrons can only be generated in pairs. Nevertheless, in our universe today there are lots of electrons and almost no positrons. The symmetry between matter and antimatter cannot be as perfect in reality as the diagrams suggest. QED alone cannot explain this – more advanced theories are needed here, and even today this question has not yet been fully clarified. If you would like to know more about the extent of this asymmetry, you can find some information in Infobox 3.1.

Infobox 3.1: The tiny asymmetry between matter and antimatter

In the universe today there exists practically no antimatter. So during the Big Bang, when matter and antimatter where created from the available energy, there must have been a slight imbalance between matter and antimatter, so that an excess of matter was left over when matter and antimatter finally annihilated each other. So how large was this excess of matter over antimatter shortly after the Big Bang?

Immediately after the Big Bang, the then tiny universe would have been filled with an extremely dense hot plasma of different particles and antiparticles, which would have constantly been transforming into one another, decaying, then reappearing again from the abundant energy available. Photons, protons, electrons, and neutrons would have been roughly equally common along with their anti-particles.

As the universe expanded and cooled down, particles and antiparticles would have annihilated each other, leaving a small particle surplus. The photons around at that time still exist today, since the photon has no anti-particle and these photons would therefore have remained intact. They still penetrate the entire present universe as a weak cosmic microwave background radiation, with a density of about 400 photons per cubic centimeter. Nowadays, these photons are about one billion (10^9) times more common than the remaining electrons, protons, or neutrons, while shortly after the Big Bang they would have been about as frequent as these particles and their antiparticles.

Thus, in the Big Bang there must have been an excess of about one electron per billion electron–positron pairs (see Fig. 3.11), and the situation would have been analogous for protons and neutrons and their anti-particles. In this sense, the asymmetry between matter and antimatter amounts to about one billionth. This is tiny, but without this imbalance we would not exist today. This small asymmetry, also known as *CP-violation*, is being intensively investigated with the help of modern particle accelerators.

Antiparticles Solve the Problem of Negative Energies

The idea that electrons can travel backwards in time has another important consequence: it solves the problem of negative energies without the need for a

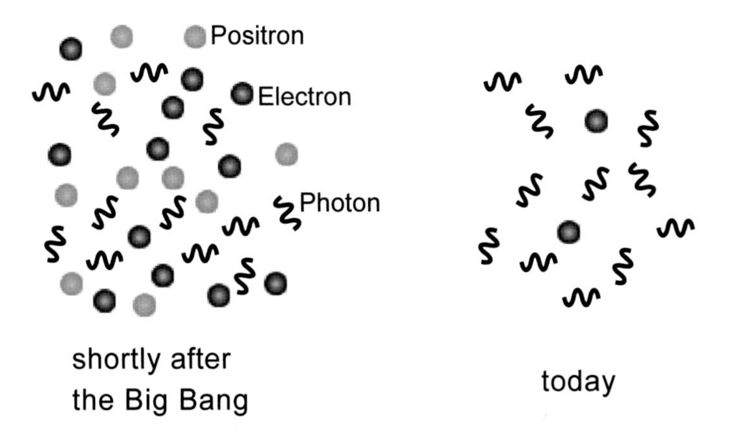

Fig. 3.11 Shortly after the Big Bang there were about the same number of photons, electrons, and positrons, with a tiny excess of about one electron to one billion electron–positron pairs. Today, now that these pairs have annihilated each other, only the tiny surplus remains, while almost all the photons have survived to become today more than a billion times more frequent than electrons (or indeed protons and neutrons)

filled Dirac sea. Instead of demanding that all negative energy states are occupied, we require that the negative energy states are exactly those that move backwards in time, while the positive energy states move forward in time as usual. An electron that moves backwards in time thus has a negative energy.

Why this should solve the problem with negative energies is perhaps not immediately obvious. The best way to understand it is to imagine each electron as a small energy account. We can draw energy from it – for example when it emits photons – whereby the electron loses energy, or we can deposit energy, which increases the energy of the electron.

The possibility of negative energies now means that we can draw any amount of energy from the electron's energy account, just as with an unlimited overdraft facility. We can extract more and more energy as often as we like, thus solving the energy problems of the whole world by pushing the energy account ever further into the negative. This was exactly the problem the Dirac sea was supposed to solve, because with a full Dirac sea all available overdraft facilities – even those with infinitely large sums of credit – are already exhausted. There is no room for another electron with a negative energy account, i.e., no more energy account can be overdrawn.

Our new solution without the Dirac sea looks different now. As soon as an energy account slips into the negative, it begins a journey into the past and is thus withdrawn from future access. So we can only use the overdraft facility once – after that, the electron with its energy account disappears into the fog of the past.

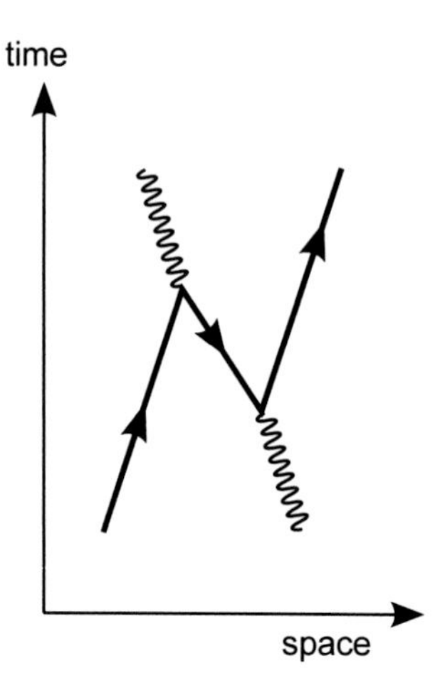

Fig. 3.12 The diagram for the scattering of light on electrons can be interpreted in two ways. Variant 1: An electron emits a photon in the upper left corner and gets into a state of negative energy due to the energy loss. This causes it to migrate backwards in time, until a photon in the lower right corner pays off the energy debts and the electron travels into the future again with positive energy. Variant 2: An incoming photon creates an electron–positron pair at the bottom right. The positron then meets another electron at the upper left corner and they annihilate each other, whereupon their energy is radiated as a new photon

Someone back there could now pay exactly the same amount of energy into the energy account that we will later take from it. In the past, this person will thus have compensated for the debts we are making today. If we look at this process in the normal time direction, we see the following. In the past, someone paid an amount of energy into the energy account, and we are withdrawing it today. It is as if credit had been deposited into the account in the past, and we are withdrawing it today. In other words, debts that travel to the past act in the same way as credit that proceeds in the normal time direction (Fig. 3.12).

This completes our picture: electrons that travel with negative energy into the past act like positrons that migrate into the future with positive energy. In this way, there is no longer any problem with negative energy states – electrons and positrons always travel in the normal time direction with positive energy. Feynman got the same results in his calculations as other people who juggled with holes in the Dirac sea, which was much more complicated. Feynman had never liked the Dirac sea anyway, owing to the logical difficulties it involved, as he reported in his Nobel Prize lecture: "I got so confused that I remembered Wheeler's old idea about the positron being, maybe, the electron going backward in time."

Give Stückelberg Back His Notes!

By the way, Feynman and Wheeler were not the only ones who came up with this idea. The Swiss mathematician and physicist Ernst Carl Gerlach Stückelberg had already proposed it in 1941, although Feynman was probably unaware of this at the time. Stückelberg also anticipated many other ideas about QED, for which Feynman, Shin'ichirō Tomonaga, and Julian Seymour Schwinger won the Nobel Prize in 1965 (see below). Some even say that Stückelberg was years ahead of his time. However, on the rare occasions when he published his work, it was in journals that were not widely read, and he did not do much to promote them, so they were largely ignored and long underestimated.

Apparently, Feynman heard about what Stückelberg had achieved in later years and paid respect to him. After receiving the Nobel Prize, he gave a lecture at the European Research Centre CERN near Geneva, i.e., just outside Stückelberg's front door, since Stückelberg was teaching at the University of Geneva. Of course, Stückelberg was also in the audience. While Feynman was surrounded by admirers after the lecture, Stückelberg left the hall alone, and Feynman remarked: "He (Stückelberg) did the work and walks alone toward the sunset; and, here I am, covered in all the glory, which rightfully should be his!" It is even said that, after the Nobel Prize was awarded, Schwinger once said to Feynman: "Now are you going to give Stückelberg his notes back?" Of course, this was meant as a joke, because Feynman, Schwinger and Tomonaga had developed their own ideas, and Feynman had taken them much further than Stückelberg – there are obviously times when certain ideas are simply in the air, so that several people may come up with them independently.

Feynman's Tool Box

With his space-time diagrams, Feynman had an effective tool at hand to address the problems of QED. The lines and vertices corresponding to the electrons and photons gave a vivid picture of the physical processes, and at the same time provided the mathematical guidelines for multiplying and summing the individual quantum amplitudes to determine the overall amplitude for the process. In addition, it was immediately apparent that the rules of special relativity were respected: switching from one observer to another, the only changes were to the locations and times of the vertices and the inclinations of the lines, and this in a simple way, while the basic structure of the diagrams remained unchanged.

Of course, the devil was still in the details: Feynman still had to find exact formulas for the individual amplitudes and set out the mathematical rules down to the last sign, so that the diagrams could be translated into formulas. Because of the problems involved in using his path integrals to establish a relativistic description of the electron spin, he proceeded intuitively and tried to guess the correct relativistic formulas. In his path integrals, for example, he initially limited himself to non-relativistic electrons, linked to relativistic photons, and then replaced the velocities of the electrons by the matrices from the Dirac equation to see if he could arrive at a relativistic description of the electrons. On the basis of his experience, it seemed obvious to him that this was the way to proceed, but he did not have any strict proof.

Since we always have to integrate over all possible positions and times appearing in the formulas, i.e., positions and times where every single vertex can be located, the integrals that arise become more and more complex, the more vertices a diagram contains. By then Feynman had become highly skilled when it came to solving such integrals. He gradually perfected his abilities and managed to tackle even complex integrals, where many others had failed.

He checked his results again and again by trying to reproduce what others had already calculated using the methods available at that time. In the end, everything began to fit and he had managed to put together a kind of manual to tackle the problems of quantum electrodynamics.

Feynman was so far the only one who had really mastered this collection of methods and tricks and there was still a long way to go before others would appreciate the great value of his approach. In addition, there were also the annoying infinities, which also plagued Feynman's method: whenever the diagram contained a loop, so that one could do a round trip around a set of lines and vertices in the diagram, the corresponding integral became infinitely large.

Feynman had found a way to make the integrals finite, at least temporarily. Simply put, he ascribed a small size to the electrons, as though treating it as a small charged sphere. We already know why this might work from our discussion of self-energy in Chap. 2. It is only when we let the radius of the sphere tend to zero that the self-energy becomes infinitely large. The same applies to the infinities in the Feynman diagrams. The trick here, mathematically speaking, is to introduce the small spatial extension of the electrons in such a way that the rules of special relativity are not violated. Feynman had succeeded in doing just that. At some point, however, the artificially introduced electron radius must be allowed to go back to zero and we still have to think about how to deal with the infinity.

What was still missing was a real challenge for Feynman's method, where it would be able to demonstrate its advantages. An experimental result was needed that could not be explained by the Dirac equation alone. And Feynman was lucky! With the measurement of the so-called Lamb shift and also the determination of the anomalous magnetic moment of the electron, two such results were obtained that were urgently in need of a theoretical explanation. This was the chance that Feynman and many other theorists had been waiting for.

3.3 Lamb Shift, Magnetic Moment and Renormalization

The word *atomic physics* is often used when referring to the atomic bomb, but this is misleading. Strictly speaking, the bomb should be called a nuclear weapon because its gigantic energy is released by atomic nuclei. Atomic physics, on the other hand, does not concern itself at all with the atomic nucleus – that would be nuclear physics – but rather with the electron shell of the atoms. These determine the atom's chemical behavior, for example, and involve much lower energies than those in the atomic nucleus. The main question is then: which quantum states can the electrons occupy in an atom, and what energies do these states have?

The situation is simplest for the most common element in the universe: hydrogen. Its atoms consist of a single proton in the nucleus and one electron in the shell. If we cannot understand the hydrogen atom, we certainly cannot understand any other atoms. So the question is: do we in fact understand the hydrogen atom?

Energy Levels in the Hydrogen Atom

It was one of the great successes of theoretical physics when Niels Bohr succeeded in calculating the energy levels of the electron in the hydrogen atom in 1913, although he had to make a number of ad hoc assumptions about the electron orbits. In 1926, quantum mechanics finally provided a clear justification for the energy levels: electrons do not move on orbits at all, but instead form standing electron waves, so-called orbitals, similar to the vibra-

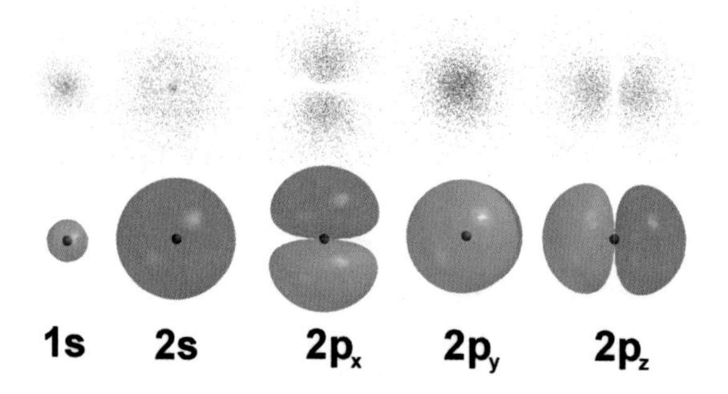

Fig. 3.13 Illustration of the standing electron waves (orbitals) of the fundamental oscillation (1s) and the first harmonics (2s, 2p) in the hydrogen atom. At the top, the probability of the electron's position is shown as a point cloud, and below this is the boundary surface at which the magnitude of the wave function falls below a certain value. Source: https://commons.wikimedia.org/wiki/File:AOs-3D-dots.png?uselang=de

tions of a guitar string, but in three dimensions. The spherically symmetric fundamental oscillation corresponds to the lowest energy level, while the various harmonics yield the higher energy values or "shells" (Fig. 3.13).

The results obtained using the Schrödinger equation were already well in line with the observed energy levels. However, looking closely enough, it could be seen that many levels actually consisted of several individual levels very close together, something known as *fine structure*. The non-relativistic Schrödinger equation could not explain this, because this is a relativistic effect in which the spin of the electron, i.e., its internal quantum angular momentum, plays an important role. The way the electron spin is oriented in the electron shell makes a small difference to the energy of the electron.

Fortunately, since 1928, the relativistic Dirac equation had been available, and that consistently took this effect into account. And indeed, when the energy levels of hydrogen were calculated using the Dirac equation, the fine structure of the energy levels could be rather accurately reproduced. Depending on how the spin of the electron was oriented, slightly different energies were obtained.

So everything seemed to be in order. Theory and experiment were in good agreement, within the limits of measurement accuracy. However, it was difficult to measure the energy levels very precisely. Here and there it seemed as if there could be small discrepancies between theory and experiment, but measurements in the 1930s were simply not accurate enough to substantiate such suspicions.

After the end of the Second World War, a technical equipment was set up that could at last tackle this problem. At Columbia University in New York

City, a stronghold of experimental atomic physics was built up under the direction of Isidor Isaac Rabi. Here physicists were able to use microwaves to measure the energy differences between closely spaced energy levels with very great accuracy.

The First Surprise: The Lamb Shift

When the American physicist Willis Eugene Lamb and his doctoral student Robert C. Retherford at Columbia University investigated the energy levels of the second shell in the hydrogen atom, they got a big surprise. Depending on the combination of orbital angular momentum and spin, there were expected to be two closely spaced energy levels in this shell – as predicted by the Dirac equation. However, what Lamb and Retherford found in April 1947 were three levels (Fig. 3.14). The lower level was further split, i.e., it contained two quantum states that did not have exactly the same energy. Their energy difference was only a tiny 4.4×10^{-6} electronvolt, where the *electronvolt* (eV) is an energy unit that is specially designed for the world of particles: if an electron passes through an electrical voltage gradient of one volt, it absorbs an energy of one electronvolt. A standard nine-volt battery can thus provide each electron with an energy of nine electronvolts, which it can release in an electric light bulb, for example.

If the electron in the hydrogen atom now jumps back and forth between the two very closely spaced energy levels, it emits or absorbs a photon with an energy of exactly 4.4×10^{-6} electronvolts. Such photons correspond to microwaves with a frequency of about 1000 MHz and a wavelength of about 28 cm, which Lamb and Retherford could measure quite accurately with their sophisticated microwave technology. Today this shift in energy is called the Lamb shift. The current measured value is 1057 MHz.

Ted Welton, Feynman's old student friend from his time at MIT, provided a vivid explanation of the Lamb shift in 1948. The electron carries a virtual photon cloud around with it all the time, and he imagined that the interaction of the electron with this cloud would lead to small statistical movements of the electron. The electron is constantly emitting virtual photons and catching them again, causing it to make tiny random jumps back and forth.

As long as the electron is at a greater distance from the proton, this has little effect on the attraction between the proton and the electron, because the random jumps of the electron are too small. However, if the electron happens to be very close to the proton, the small random jumps lead to a noticeably fluctuating attraction. If this is calculated exactly, these fluctuations cause on average a slight weakening of the attractive force close to the proton.

So how would this affect the electron waves of the second shell, the ones that are of particular interest to us when we consider the Lamb Shift? The attenuation of the attractive force close to the proton has hardly any influence on the so-called 2p states, because this quantum wave has minimal values at the position of the proton in the middle of the atom. The 2p quantum wave hardly oscillates there and the probability of the electron staying there is close to zero (Fig. 3.13). In the p states the electron rarely comes very close to the proton.

The situation is different for the spherically symmetric 2s state. Here the electron has a greater probability of being close to the proton, so the attractive force is slightly weakened by its small random jumps. The 2s state is therefore raised slightly in energy terms compared to the 2p state, as shown in Fig. 3.14. Without the small random jumps, both states would have the same energy, according to the Dirac equation.

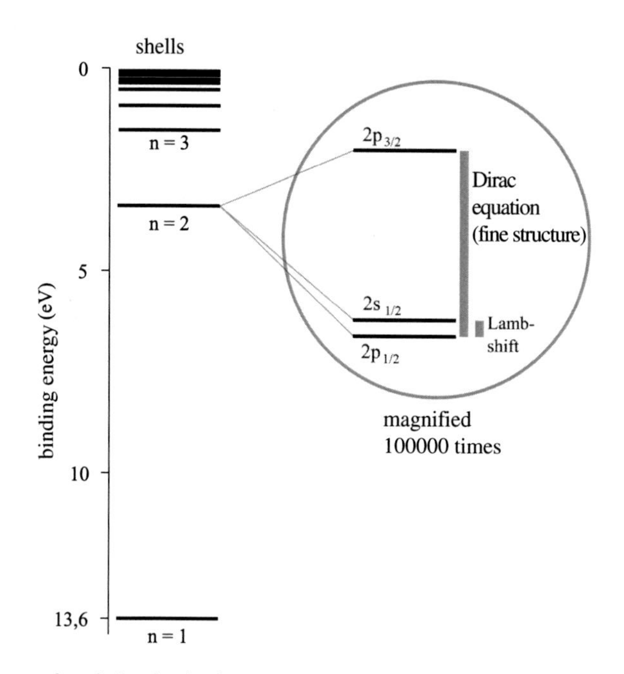

Fig. 3.14 Energy levels in the hydrogen atom. For the second shell, the splitting due to the Dirac equation and the Lamb shift is also shown. The Lamb shift causes a slight upward shift of the $2s_{1/2}$ level compared to the $2p_{1/2}$ level. The index (i.e., 1/2 or 3/2) indicates the total angular momentum of the state, which is composed of the spin and orbital angular momentum of the electron. The s levels have orbital angular momentum 0, and the p levels 1

The Second Surprise: The Magnetic Moment of the Electron

The Lamb shift was a precisely measurable physical phenomenon that could not be explained by the Dirac equation. But Rabi's group at Columbia University (Kusch, Foley, and others) found another such phenomenon: it turned out that the magnetic moment of the electron was about 0.1% larger than predicted by the Dirac equation.

The fact that the electron has a magnetic moment means that it behaves like a tiny magnetic compass needle. The reason for this lies in the spin of the electron, i.e., in its internal quantum angular momentum. If one imagines the electron as a tiny electrically charged sphere, then the charge on the surface of this sphere circles around the axis of rotation. As we know from the Maxwell equations, such an electric circuit current generates a magnetic field that is very similar to the magnetic field of a compass needle, aligned parallel to the axis of rotation (Fig. 3.15). Each electron should therefore have a magnetic moment like a tiny compass needle.

Now, this descriptive idea is by no means an absolute necessity, because in reality an electron is not a small charged sphere behaving in accordance with the laws of classical physics, but rather a point-like quantum object.

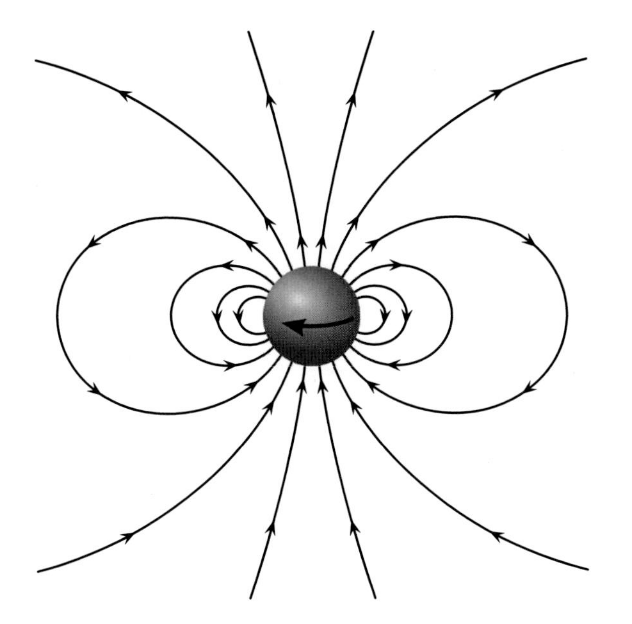

Fig. 3.15 A rotating electrically charged sphere generates a magnetic field similar to a small compass needle. The magnetic moment of the electron can be illustrated using this model

However, we can nevertheless check it out, to see if there really is something in this idea.

First of all, electrons do actually possess a magnetic moment, as predicted by our model. The magnetic field of a real compass needle is in fact created by the cooperation of many electrons, which align their magnetic moments so that they are all parallel.

It remains to determine the magnitude of the magnetic moment of the electron. Is it actually as big as we would expect from a rotating charged sphere (with a charge distribution equal to its mass distribution)? If so, the so-called g-factor is said to be equal to one, i.e., $g=1$. So how large is the g-factor of the electron?

If we believe the Dirac equation, the magnetic moment of the electron is exactly twice what one would normally expect from the spherical model: $g=2$. This is quite conceivable, because the electron is not really a sphere.

So which is right? If we measure the magnetic moment of the electron in the experiment, we actually find the value $g=2$, i.e., the Dirac equation carries the victory – at least almost. For when Rabi's group looked very closely at this quantity for the first time, they found a small deviation: g is slightly greater than 2, which is why we speak of the *anomalous magnetic moment.* Modern measurements provide the value

$$g = 2.00231930436182(52)$$

where the number in brackets indicates the uncertainty in the last two digits. It is amazing how accurately this number can be measured today! There is hardly any other measurement in physics for which similar accuracy can be achieved. If, for example, the distance from Hamburg to Munich (approximately 600 km) were determined with the same accuracy, the result would be fixed to within a hair's breadth.

The Shelter Island Conference

Could quantum electrodynamics explain the Lamb shift and the magnetic moment of the electron? This was the main topic of a small conference that took place from June 2 to 4, 1947 – only two months after the discovery of the Lamb shift – in a small hotel on Shelter Island at the eastern end of Long Island in the state of New York. After the Second World War, this was the first opportunity for a number of young American physicists to discuss the basics of quantum theory.

With only 24 physicists present, the number of participants was fairly small. This was quite deliberate, as it would allow intensive discussions between the participants. Robert Oppenheimer, Hans Bethe, and John Archibald Wheeler were also present in addition to Richard Feynman and another very talented young physicist who will receive more attention in the next section: Julian Schwinger.

Feynman later recalled that the Shelter Island Conference was the most important conference he had ever attended. Here, Willis Lamb and Isidor Rabi presented their experimental results on the Lamb shift and the *g*-factor of the electron, thus making it clear for the first time that the Dirac equation was incomplete. There had to be small corrections, and these seemed to result from the same effects in QED that caused the disturbing infinities.

After almost two decades of stagnation and confusion, Lamb's and Rabi's experiments finally came to bear on the deadlock in quantum electrodynamics. For the first time there were clear data which could not be explained by the Dirac equation alone and which had to be computed! To achieve this, the infinitely large corrections that QED delivered had to be reduced somehow to small ones, because the Lamb Shift was only a tiny effect. The experiments showed that this had to be possible if QED was to make sense as a theory – but how could they do it?

Several people such as Hendrik Kramers and Victor Weisskopf, who also took part in the conference, had already come up with ideas to solve this problem. The crucial insight was that the mass and charge of the electron, which were used in the formulae as mathematical parameters m and e, could not simply be identified with the physical mass and charge of the electron. Rather, the parameters m and e were the *naked* or *bare* values of the mass and charge, before taking into account any QED corrections due to the virtual photon cloud and other virtual particles around the electron. On the other hand, the cloud of virtual particles could not be neglected – it belonged inseparably to the physically observed electron. Only the electron together with its virtual particle cloud (Fig. 3.16) has the mass and charge that we actually observe in experiments, while the bare electron is a theoretical construct used for the mathematical formulation of the theory, so there is no reason why its bare mass and charge should not approach zero or infinity to compensate the infinity of the virtual photon cloud. We may put it like this: the infinities can be hidden in the physical charge and mass of the electron, and it is assumed that they are compensated there by a suitable bare charge and bare mass.

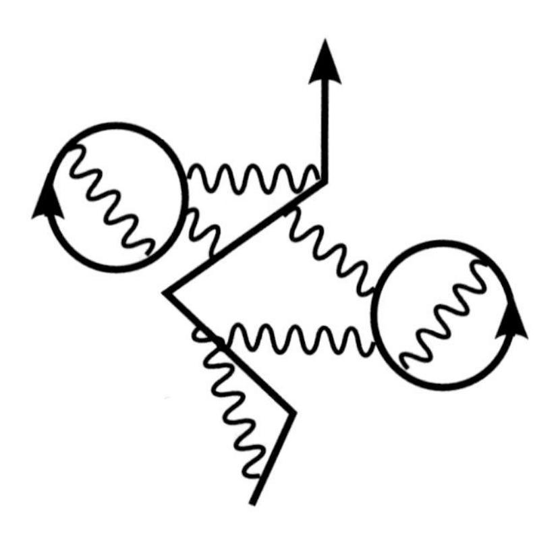

Fig. 3.16 In a physical electron, the bare electron (solid line) is inseparably linked to a cloud of virtual particles (photons and electron–positron pairs)

Admittedly, this process, known as renormalization, which we will be going into in more detail below, may give us an uneasy feeling since the corrections are infinitely large. The infinitely large corrections must be compensated in such a way that something small remains. Moreover, this must be done in a clearly defined way, i.e., in spite of the infinities, the process of renormalization must not be arbitrary. There must be a clear result, and this result should coincide with experimental measurement. Would QED pass this test?

Hans Bethe Calculates the Lamb Shift – At Least Non-relativistically

After the conference, Hans Bethe was the first to work out the energy shift in the hydrogen atom measured by Lamb. He was convinced that, if there was such a clearly measurable deviation from the Dirac equation, then every effort had to be made to calculate it. For Bethe, understanding physics meant being able to calculate a definite number. If you couldn't do that, you didn't understand the physics.

Bethe did not know how to describe the electron in QED in a relativistically correct way, but he was well versed in the sophisticated methods of the non-relativistic Schrödinger equation and knew how to describe radiation processes with relativistic photons. After the conference, when he left Shelter Island by train for Schenectady – where he worked for a while as a consultant for General Electric – he started working on the problem and at the end of the four-hour train journey, he had actually calculated a definite number for the energy shift: 1040 MHz. This was very close to the experimental result and Bethe took it as a great success. The news spread quickly. In fact, it was in a sense almost too good, because his calculations contained a somewhat questionable assumption which he had to make in order to get the infinity under control. More precisely, he assumed that there had to be an upper limit for the energy of the virtual photons, approximately equal to the electron mass (expressed in units of energy) – see Infobox 3.2 for details.

Infobox 3.2: Hans Bethe's calculation of the Lamb shift

How did Bethe calculate the Lamb shift? To begin with, he simply used the non-relativistic Schrödinger equation to describe the electron in the hydrogen atom, which is usually sufficient. Then he added the influence of the photon cloud on the electron as a small disturbance. This perturbation, however, makes an infinite contribution to the energy of the electron in the hydrogen atom. On the other hand, the photon cloud also makes an infinite contribution to the energy of a free electron.

What we measure experimentally is the binding energy of the electron in the hydrogen atom. Thus, we compare the energy an electron has in the electric field of the proton with the energy it has without this field. Bethe therefore subtracted the energy of a free electron from the energy of an electron in the hydrogen atom, both calculated with the contribution of the photon cloud – this is the central idea of renormalization.

Both energies become infinitely large due to the contribution of the photon cloud, so the result is effectively infinity minus infinity, and mathematically speaking, this can be absolutely anything. Therefore, we have to be a little more subtle and look at where the infinities come from. They arise from the fact that the virtual photons can have all possible energy values, whereby there is no upper limit, and the fact that all these energies have to be summed up.

The trick now is to introduce an artificial upper limit K for the photon energies. With this limit K we can now proceed as we would with a finite number, thereby keeping the infinities under control. In technical terms, this process is also called *regularization* of the infinity; it is an important step in renormalization.

What happens when we calculate the difference between the energy of the free electron and the energy of the electron in the hydrogen atom? Both energies include terms with this artificial upper limit K, and these terms become infinitely large when K is allowed to grow to infinity. However, when we calculate the difference between the two energies, the terms with K may possibly cancel each other. In this case, we may subsequently let K grow towards infinity and the energy difference will remain finite, because it no longer contains K.

This was indeed the case in Bethe's calculations, for those terms that go towards infinity at the fastest rate when K increases. So the most dangerous infinities in the energy difference cancel each other as desired.

However, Bethe's idea didn't work perfectly, because a term with K was left in the energy difference. This term essentially corresponds to the natural logarithm of K, when K is given in electronvolt (eV). So Bethe hadn't really managed to get rid of all the infinities. How should he cope with the annoying logarithm of K in the energy difference?

Bethe argued that, in a completely relativistic calculation, a finite result was to be expected.[2] His non-relativistic calculation, on the other hand, should only be taken seriously up to photon energies that are not yet sufficient to produce new electrons, because then we must take the theory of relativity into account. So he simply used the electron mass expressed in the energy unit eV, i.e., about 500,000 eV, for the upper limit K, and thereby made the logarithm of K finite. This trick actually worked, and Bethe got a fairly good value for the Lamb shift, even though his argument for the upper limit of the virtual photon energy was a little shaky.

Here Bethe benefitted from the fact that the logarithm of K only grows very slowly with increasing K, so we are in fact dealing with a very mild infinity. This is also typical for the infinities in QED. In particular, the logarithm of K is rather insensitive to the exact value of K as long as it remains at a certain order of magnitude. For example, $\ln(500{,}000) = 13.12$ and $\ln(1{,}000{,}000) = 13.81$ are close to each other, although in the second case we take the logarithm of a number twice as large – try it out with your calculator! So it did not matter so much which number Bethe used for the upper limit K in his calculation.

Immediately after his arrival in Schenectady, Bethe called Feynman and told him excitedly about his result and how he had used the idea of renormalization to get the infinity at least partially under control. Feynman didn't fully understand how important Bethes result was for QED, but Bethe was so excited that it seemed to be something really significant.

Back in Ithaca, Bethe then gave a lecture and explained in detail how he had calculated the Lamb Shift approximately using the non-relativistic Schrödinger equation. He expressed the hope that his arbitrary assumption of a photon energy cutoff would not be necessary in a fully relativistic calculation.

The Relativistic Lamb Shift – "I Can Do It!"

Now relativistic calculations were Feynman's specialty, because his path integrals and Feynman diagrams automatically took relativistic rules into account. So after the lecture he went to Hans Bethe and confidently proclaimed: "I can do that for you. I'll bring it over tomorrow!"

This would turn out to be a little over-optimistic. Until that time Feynman had been mainly concerned with the special form of electrodynamics without self-energy that Wheeler and he were still trying to turn

[2]H. A. Bethe: *The Electromagnetic Shift of Energy Levels*, Phys. Rev., 72, 339 (1947)

into a quantum theory. But now it was a question of looking at the standard form of electrodynamics, because the Lamb shift was apparently caused by the self-energy of the electron, i.e., the influence of the virtual photon cloud. However, Feynman did not yet have a complete toolbox to express and calculate QED with his diagrams – this would only emerge gradually.

In addition, Feynman had little experience in calculating specific quantities like the self-energy. So the next day he went to Bethe and asked him if he could teach him how to do it. Bethe duly showed him how he would do things, and Feynman tried to translate Bethe's non-relativistic methods into his own relativistic language. It didn't work right away, but bit by bit Feynman learned how to do such calculations using his own methods, and he eventually began to see how renormalization worked. He found that his own approach was much more efficient than the other methods available to date – many things worked out more easily, while his colleagues had to carry out lengthy calculations involving many terms. After a few weeks he had succeeded in calculating the Lamb shift without having to rely on cutting off the higher photon energies as Bethe had had to do.

The Feynman diagrams up to the second order in perturbation theory are shown in Fig. 3.17. In first order (a), a photon couples directly to the atomic

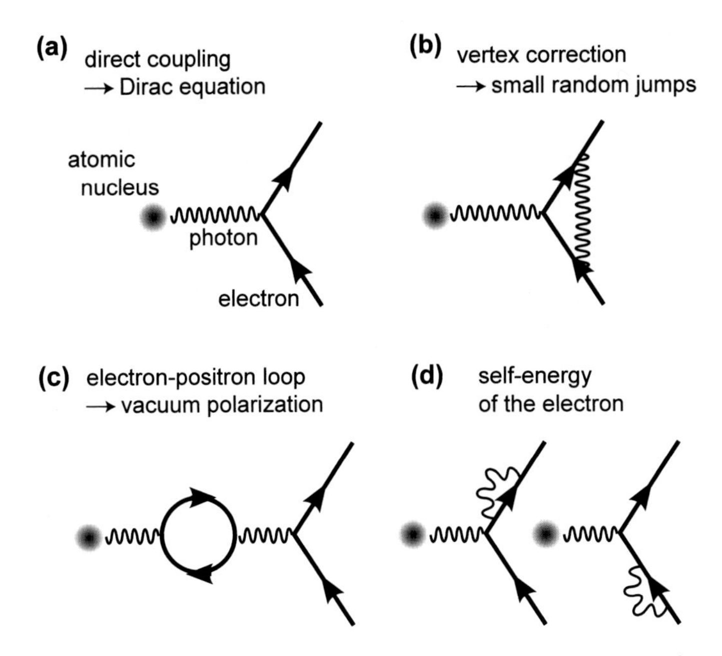

Fig. 3.17 Feynman diagrams for the interaction between a proton and an electron in first order (a) and second order (b)–(d)

nucleus and the electron. If only this diagram were taken into account, we would obtain exactly the same result as with the Dirac equation, and there would be no Lamb shift and no anomaly in the magnetic moment. So diagrams with more vertices have to be included. In second order, these are the diagrams (b), (c), and (d),

Diagram (b) is the most important here. The electron emits a virtual photon and thus jumps a little to the side before it couples to the photon from the atomic nucleus and then recaptures its virtual photon. This produces the tiny random jumps of the electron with which Ted Welton had so vividly explained the Lamb shift, because the electron sees a somewhat shaky and thus weakened field in the vicinity of the atomic nucleus.

The two diagrams in (d) are less important here, although for reasons of completeness they must be taken into account for renormalization, since they have an influence on the self-energy of the electron. Another interesting diagram is (c). It shows that the virtual photon coming from the proton temporarily transforms into a virtual electron–positron pair before it reaches the electron. This expresses the idea that the proton is surrounded by a cloud of virtual electron–positron pairs, and we can imagine that the virtual electrons buzz more closely around the proton than the virtual positrons because of the electrical attraction. As a result, the vacuum behaves like a polarizable medium and this effect is therefore called vacuum polarization (Fig. 3.18).

When the electron is particularly close to the proton, it penetrates the shielding virtual electrons in its vicinity and therefore perceives a stronger proton charge than it would further away from the proton. The polarization cloud of diagram (c) thus increases the attractive force on the electron very close to the proton and therefore has exactly the opposite effect to the

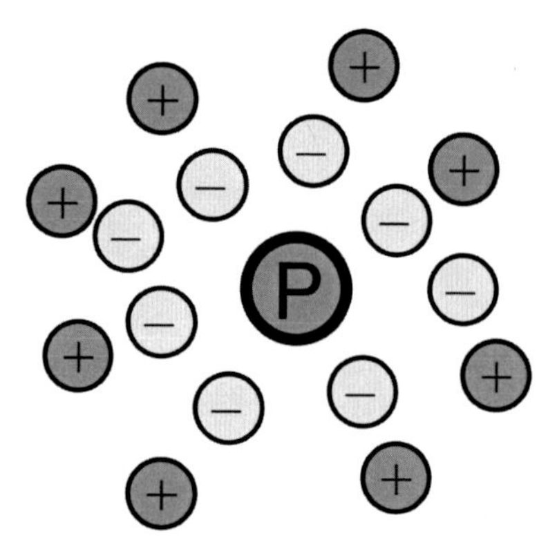

Fig. 3.18 The loop diagram (c) leads to a polarized cloud of virtual electron–positron pairs near the proton (vacuum polarization)

random jumps of diagram (b). If there were only the vacuum polarization, the $2s_{1/2}$ level would be lowered by 27 MHz compared to the $2p_{1/2}$ level, instead of being raised by 1057 MHz – the Lamb shift would go in exactly the opposite direction. Apparently, the random jumps of diagram (b) are more important than the vacuum polarization of diagram (c). But only if the small opposite effect due to vacuum polarization is taken into account will the QED calculations agree with experiment.

Feynman did not particularly like the loop diagram (c) and ignored it in his first calculations, because it did not fit with his idea that only one electron should move through his diagrams. So there he was again, the rather too headstrong Feynman! Later he realized that he could not do without the loop diagram, because the theory would otherwise be inconsistent – for example, there are problems with the probabilities, which have to add up to 100%. But even in his famous QED paper *Space-Time Approach to Quantum Electrodynamics* from 1949, he remained somewhat sceptical. In a footnote he wrote that it would be very interesting to calculate the Lamb shift accurately enough to be sure that the 20 MHz expected from vacuum polarization was actually present. Feynman's wish was later fulfilled: vacuum polarization is indeed real!

The Tamed Infinity

Since diagrams (b)–(d) each involve a loop which we can run round in a circle, they each make an infinite contribution. This infinity arises as the vertices move ever closer together and the loop becomes ever smaller. Since we must integrate over all vertex positions in space and time, these increasingly smaller loops also appear in the integrals and must be considered. This is exactly what leads to the annoying infinity. The shrinking loops correspond to ever shorter wavelengths and thus ever higher particle energies within the loop.

To deal with the infinities, we must first make them finite, i.e., regularize them, by introducing a kind of minimum distance for the vertices – or equivalently, a maximum particle energy, as in Bethe's calculation of the Lamb shift. Feynman did this in a very clever way, so that his regularization was compatible with the theory of relativity. However, his regularization method did not work for the vacuum polarization loop diagram (c) that he so much disliked. Feynman struggled with this problem for some time, but was unable to find a satisfactory solution. It was only when Hans Bethe told him at the end of 1949 about a method that Wolfgang Pauli had discovered, that the infinity due to this diagram could finally be tamed.

For the Lamb shift and the *g*-factor, the diagrams provide a sum of mathematical terms, some of which become infinite if the minimum separation of the vertices is allowed to approach zero, or equivalently if the maximum energy of the particles within the loops is allowed to go to infinity. In the next step, these infinities are hidden in the physical charge and mass of the electron, under the assumption that they are compensated there by a suitable bare charge and mass – this is precisely the famous renormalization procedure, which we have already mentioned several times. You can read more about this topic in Infobox 3.3.

Infobox 3.3: A renormalization to try out for yourself

Dealing with infinities in renormalization often seems rather confusing. How can we calculate something meaningful when infinities are involved? The following example is designed to shed some light on this.

Suppose we are dealing with a particle associated with some quantity E which we measure at a distance x from the particle – it could, for example, be the physical charge of the particle, seen from this distance. Suppose our theory were to give the following formula for E:

$$E = \frac{e}{1 - e(x + K)}.$$

This is not a real formula from QED, but a simple formula to be taken as an example, which is intended to illustrate the essentials of renormalization in a simplified way. In this formula e stands for the bare charge of the electron and the symbol K stands for quantity that tends to infinity, as in Bethe's calculation, i.e., we must imagine that we will let K become infinitely large.

Does this formula make any sense? If we use a specific number for the bare charge e and let K become infinitely large, our quantity E becomes infinitely small and thus ultimately zero, regardless of what distance x we may choose. This would be pointless for a finite measurable quantity. Furthermore, we would not know what number to use for the bare charge e, because e is not measurable.

Can the problem be cured? Assume we know from experiments that at a certain distance x_0 the measured quantity E has the value E_0. Then, according to the above formula, we should have:

$$E_0 = \frac{e}{1 - e(x_0 + K)}.$$

After a little manipulation, we can calculate the bare charge e from this formula:

$$e = \frac{E_0}{1 + E_0(x_0 + K)}.$$

In the above formula for the quantity E, we can now insert this expression for the bare charge e and after a little more manipulation, obtain

$$E = \frac{E_0}{1 - E_0(x - x_0)}.$$

The calculation is not difficult – try it out for yourself on a piece of paper! You will see how the infinity K disappears. The new formula for E contains neither the infinity K nor the non-measurable bare charge e. Instead, it contains the measured value E_0 at the distance x_0, so we can now calculate the quantity E for every other distance x using this formula.

So in our formula we were able to eliminate the infinity K and the bare charge e with the help of a measured value and thus make sense of the formula. This is not possible with all formulas. The above formula can be renormalized in this sense.

In general, theories are called renormalizable if we can in fact get rid of all infinities by using a finite number of measured quantities. QED is an example of such a renormalizable theory – the measured quantities here are the physical charge and mass of the electron.

In this way all infinities can be systematically eliminated to obtain a finite expression for physical quantities such as the Lamb shift or the *g*-factor. We also say that QED can be *renormalized*. This is not a matter of course – it only works by virtue of the special structure of the electromagnetic interaction. This method is also successful in the quantum theories of the strong and weak interaction, but not in gravitation, as we shall see later.

Feynman was not particularly happy with this way of getting rid of the infinities. We must not forget that his original goal had been to eliminate the infinities at the outset by completely banning the self-energy of electrons from the theory. He and Wheeler had been quite successful with this idea in classical electrodynamics, but they had not made any progress in quantum theory. Feynman was forced to allow self-interaction again and was then able to calculate the Lamb shift using the renormalization method.

In his Nobel Prize speech many years later, Feynman said that he still thought that renormalization theory was simply a way to sweep the difficulties associated with divergences in electrodynamics under the rug. For him, the infinities were a sign that QED was not a consistent theory in itself.

Modern research confirms Feynman's intuition to a certain extent. There are indications that QED gets problems despite renormalization when diagrams with more and more vertices are included and when higher and higher energies are considered.

So QED is probably not a perfect theory. Today we can take a little more relaxed view of this than in Feynman's days, because we now know that QED does not have to be, and probably cannot be, a perfect theory. Rather, QED is regarded as a theory that only approximately describes real physics at very short distances, i.e., at very high energies.

Where would the minimum distance or maximum energy be, at which we would expect deviations from QED? Different answers can be given, depending on which particles and interactions are taken into consideration. However, there is a final limit, beyond which even the usual conception of space and time as a continuous structure will most likely collapse, and shorter distances will be physically meaningless. To determine this limit, we have to consider what happens when quantum theory meets gravity, the weakest of all interactions: the shorter the distances we wish to look at, the shorter the wavelength of the corresponding quantum waves must be, and the higher the energy of the corresponding particles. Now, according to Einstein, every energy corresponds to a mass, so energy also exerts a gravitational force. At extremely high energies and thus short wavelengths, there will at some point be so much energy concentrated in the particle that its gravitational force will create a mini black hole with the size of the wavelength. Smaller distances can then no longer be resolved, because a further reduction of the wavelength would lead to an enlargement of the mini black hole, and it would be impossible to see anything within it. Physics on these length scales must be completely different from everything we know today, because quantum waves and billowing mini black holes will be so closely interwoven there that this will change the very structure of space and time.

The length scale where this happens is called the *Planck length*, and the associated energy the *Planck energy*. The Planck length is 1.6×10^{-35} m, twenty orders of magnitude smaller than a proton. In other words, the Planck length is in the same relation to a proton as a proton to a distance of about 150 km. Thus, the Planck length is well below all length scales experimentally accessible to us today.

The same can be said of the associated Planck energy, which is 1.2×10^{19} GeV (giga-electronvolt). For comparison, the maximum energy of a proton collision at the world's largest accelerator, the Large Hadron Collider (LHC) at CERN near Geneva, is 13,000 GeV, which is about 15 orders of magnitude below the Planck energy.

QED cannot therefore be a universally valid theory up to arbitrarily high energies. At energies available today, on the other hand, it is as accurate as we could wish for. In principle, everything relating to electrical charges and photons can be calculated with this theory. For example, QED can now be used to calculate the *g*-factor of the electron to an accuracy of about ten decimal places in very complex calculations including many Feynman diagrams. This number corresponds digit by digit to the value determined experimentally. Few physical theories have been as successful as QED!

3.4 Schwinger, Tomonaga, and Dyson

Feynman was not the only one who managed to carry out a relativistic calculation of the Lamb shift after the conference on Shelter Island. Others were also able to calculate relativistically, albeit using more cumbersome methods which were less obviously compatible with relativity theory than Feynman's approach.

Julian Schwinger – The Perfectionist of the Abstract

Julian Schwinger probably had the greatest influence among these physicists (Fig. 3.19). He was as young as Feynman – only three months older – and had already achieved a certain level of fame in physics, while Feynman was still relatively unknown. Like Feynman, Schwinger was also very talented at a young age and learned a lot on his own. Both tended to follow their own mind and didn't like being told what to do. Unlike Feynman, Schwinger had many successful students during his scientific career, including several future Nobel Prize winners.

In later years, both Schwinger and Feynman were well-known for their legendary lectures, although their style differed considerably: while Feynman cultivated a rather relaxed, entertaining style, Schwinger seemed almost to be celebrating his lectures. They were silky smooth and structured to perfection, like a symphony. This was very impressive, but at the same time could be intimidating for his audience.

Like Feynman, Schwinger had an excellent knowledge of quantum electrodynamics. While the impulsive Feynman approached the associated problems rather intuitively, and tended to move toward to the solution with great leaps, filling the gaps only later, the invariably polite Schwinger always pro-

Fig. 3.19 Julian Seymour Schwinger (1918–1994). (© Emilio Segre Visual Archives/American Institute of Physics/Science Photo Library)

ceeded in a rather structured and logical way. Schwinger was a perfectionist and loved to formulate his theories mathematically as generally and elegantly as possible. He was a master at handling mathematically abstract formalisms, and with considerable virtuosity, but often lost his audience, so that they had the impression that only Schwinger himself was able to master the extensive mathematical apparatus he had exposed. It was tempting to trust his theories, but without actually understanding them.

Schwinger and Feynman had already met briefly in Los Alamos in the summer of 1945, but afterwards their paths separated. Only at the conference on Shelter Island did they meet again to hear first-hand about the measurement of the Lamb shift and the *g*-factor.

Shortly after the conference, Julian Schwinger married his wife Clarice and they went on their honeymoon for the next two months. This time-out from physics was perhaps exactly what Schwinger needed to recharge his batteries, because afterwards he returned with great energy to the questions that the Shelter Island conference had raised: was it possible to calculate the *g*-factor and the Lamb shift in QED, or would QED fail because of its infinities?

Calculating the g-Factor

The calculation of the Lamb shift by Hans Bethe had already shown shortly after the conference that it should be possible. Schwinger set to work and actually succeeded in calculating the magnetic moment of the electron – the *g*-factor – to a good approximation by the end of 1947. Like Feynman, he used the ideas of renormalization to get rid of the infinities. The formula he obtained in second order perturbation theory is quite simple. It reads

$$g = 2 \cdot \left(1 + \frac{\alpha}{2\pi}\right),$$

where $\alpha \approx 1/137$ is the fine-structure constant, i.e., the squared charge of the electron in natural units, which we have already encountered.

The fact that the correction term is proportional to α should come as no surprise. The corresponding second-order diagrams have two vertices more than the first-order diagram, and each vertex contributes a factor $\sqrt{\alpha}$ to the amplitude. Since α is somewhat fortunately a rather small number, the corrections become smaller and smaller the more vertices there are in the corresponding diagrams. So the above second order approximation formula should already give quite a good result.

If we insert the number 1/137 for α in the formula, we get

$$g = 2.002\,323\,4\ldots$$

which in fact already matches the experimental value g = 2.002 319 304 361 82(52) quite well. It was a huge success for Schwinger, who was the first to provide a correct calculation of the simplest deviation term $\alpha/(2\pi)$ from the result of the Dirac equation. This famous term now appears on his gravestone, which he shares with Clarice, his wife for 47 years (Fig. 3.20).

Schwinger was equally successful with the Lamb shift. At the annual meeting of the American Physical Society (APS) in New York in January 1948, he presented his results to a large audience – it was a great triumph for him. Practically everyone present wanted to hear his lecture, and it had to be repeated in the afternoon. Feynman was also there. He had obtained the same results as Schwinger with his methods, but when he announced this after Schwinger's lecture, the reaction was not enthusiastic – it probably sounded too much like the exclamation of a little boy saying: "I can do it too!" But Feynman had only wanted to make it clear that Schwinger's results had to be correct, since he had obtained the same. So both methods seemed to give correct results, albeit in very different ways.

Fig. 3.20 The gravestone of Julian and Clarice Schwinger (© Jacob Bourjaily; CC-BY-SA 3.0 Unported; https://en.wikipedia.org/wiki/File:Julian_Schwinger_headstone.JPG)

For Feynman this realization was very important, because he now knew that he was also on the right track.

Unfortunately, hardly anyone took Feynman and his rather intuitive methods seriously at that time, while Schwinger was well known and people trusted his complicated calculations, since they were ultimately based on the proven methods of quantum field theory. However, hardly anyone was able to follow the details of Schwinger's obscure calculations. As Freeman Dyson once put it: "Schwinger did not explain how to do it, but that only he could do it".

The Pocono Conference – Schwinger's Triumph and Feynman's Disaster

Feynman's first opportunity to explain his unusual approach to QED came about three months later at the Pocono Conference, which took place in early April 1948 at the Pocono Manor Inn in northeastern Pennsylvania, roughly halfway between New York and Ithaca. This conference followed on from the previous year's Shelter Island Conference, where Rabi and Lamb had announced their experimental results on the Lamb shift and the g-factor. This time, too, the number of participants was kept relatively small, with just 28 physicists, because it was hoped to generate a fruitful discussion on the latest developments in quantum theory. Eugene Wigner, Paul Dirac, and Niels Bohr, i.e., some of the famous minds that had co-founded quantum theory, were also present in the mountains of Pocono. What would they say about the contributions of Schwinger and Feynman?

It was Schwinger who spoke first. He gave a truly marathon lecture in which he explained in detail his mathematically ingenious approach to QED, showing how the Lamb shift and the g-factor could be calculated and how the results could be renormalized. As usual, Schwinger's lecture was fine-tuned down to the last detail, and hardly anyone dared ask any questions, so as not to interrupt the wonderful flow of thoughts, as Feynman later recalled. But the more experienced veterans in the audience like Bohr, Dirac, and Teller were not so easily intimidated, and finally the first questions came up. As so often happened, it was difficult to follow Schwinger's complex lines of thought. In the end the session lasted well into the afternoon.

Schwinger's lecture made a big impression on the audience, and some of the participants, including Enrico Fermi, wrote down all the details. Back in Chicago, Fermi, Teller, Wentzel, Yang, and others spent several weeks together trying to figure out what was in their notes. They did not really succeed – the work of the young master Schwinger remained partly indigestible to them, even though it was clear to all that he had quite definitely achieved something very important.

After Schwinger's lecture it was Feynman's turn – his lecture was entitled *Alternative formulation of quantum electrodynamics*. Bethe had observed that Schwinger was least often interrupted when he argued mathematically rather than physically, so he advised Feynman to do the same. What had worked well for Schwinger, however, was exactly the wrong way for Feynman – on the contrary, he was used to bringing the physical view into the foreground. His method was based on intuition, with which he guessed the right formulas on the basis of pictures of the physical process. In contrast to Schwinger, however, he had no mathematical framework from which his theory could be rigorously derived.

So the talk didn't go very well for Feynman. Moreover, his listeners were probably already quite exhausted from Schwinger's lecture and had some difficulties getting into Feynman's unfamiliar ways of thinking. As a consequence, he was interrupted again and again. They asked him how he knew that his formulas were correct, how he had derived them, and so on. Feynman could only answer that his method gave the right results – he knew that, because he had spent a good deal of time talking with Schwinger at the conference, during the breaks between meetings. They had compared their calculations, finding that the results were consistent, even though neither of them understood the other's methods.

Dirac asked if Feynman's theory was *unitary*, whereupon he had to explain to Feynman what that meant – effectively, that the overall probability that something happens is unity. Feynman could not answer the question because he hadn't yet given it any thought. As we already know, Dirac's question was quite justified, because it is only if we consider the electron–positron loop diagram that everything fits together, and Feynman had not yet done that.

By the time Feynman came to speak about the paths of particles as they appeared in his path integrals and diagrams, some of his listeners had reached the end of their patience. Niels Bohr stood up and announced that it had been known since 1925 that the classical path of a particle in quantum mechanics was not an observable quantity and could therefore play no role. That was true – in quantum theory, a particle does not actually move along any certain path – but Feynman hadn't claimed that at all. On the contrary, every possible path contributes to the total quantum mechanical amplitude. Obviously Bohr and others had not understood that during his lecture. In short, Feynman had not succeeded in explaining his ideas in an understandable way. "My machines came from too far away – it was a hopeless presentation," he recalls later.

Feynman Won't Give up

Someone else might have been discouraged and given up after such a disaster, but that was not Feynman's way. He now knew what he had to do: he had to bite the bullet and accurately write down and publish his theory – a task he had always hated, as we know. But only in this way would other physicists have the opportunity to look at and understand everything step by step.

For this purpose, Feynman worked once again through his methods and tested them on a wide variety of problems. Step by step he perfected them and finally created a very effective toolbox that nobody else had available to them at that time. How superior his methods actually were became clear to him at the next major meeting of the American Physical Society (APS) in New York in January 1949. There, a young physicist named Murray Slotnick presented the results of a calculation which had taken him six months. Using the classical methods of quantum field theory, he had calculated how an electron and a neutron interact when so-called mesons are exchanged between them instead of photons. Mesons were on everyone's lips at the time, because it had recently become possible to produce large numbers of these short-lived particles at the newly built accelerators. Today we know that mesons consist of a quark and an antiquark, but of course no one knew that at the time.

Oppenheimer was among the audience and questioned Slotnick's results in his typical rather brusque manner. Now Feynman's curiosity was aroused: overnight he had calculated what Slotnick had taken months to do and compared his result the following day with Slotnick's. They agreed, and moreover, Feynman's calculation was even more general. Slotnick was shocked: "What do you mean you worked it out last night – it took me six months!"

That was a thrilling moment for Feynman. Apparently, he was the only one who was able to perform complex calculations in quantum field theory within a few hours. "That was my moment of triumph in which I realized I really had succeeded in working out something worthwhile", he says in his Nobel Prize speech.

The other conference participants also noticed what Feynman had achieved. Now, at last, everyone urged him to publish his methods. In April and May 1949 – about a year after the Pocono Conference – Feynman finally submitted two papers to the renowned journal *Physical Review*, and they were published there in September 1949. In the first of them – *The Theory of Positrons* – he shows how positrons can be interpreted as electrons running backwards in time, while in the second publication *A Space-Time Approach to QED*, he explains how all processes in QED can be calculated using his diagrams.

Feynman's long delay in publishing was also due to his uncertainty over how to explain his methods conclusively. He had no solid theoretical foundation from which these tools could be derived in a straightforward manner, and this repeatedly prompted others to criticize his intuitive method. How could one really be sure that Feynman's method was coherent in itself? "Well, so what", he must have said to himself finally – in his Nobel Prize speech he says: "Nevertheless, a very great deal more truth can become known than can be proven."

Freeman Dyson – It Doesn't Work Without Mathematics

However, someone else was not satisfied with this pragmatic view: the young British physicist and mathematician Freeman Dyson (Fig. 3.21). In autumn 1947 he came as a graduate student from Great Britain to Cornell University in the USA in order to be as close as possible to current developments in physics, and – like Feynman – to work closely with Hans Bethe. Bethe was enthusiastic about Dyson's extraordinary talent and gave him as a

Fig. 3.21 Freeman Dyson at Harvard University in May 2004 (© Lumidek, CC-BY-SA 3.0 Unported; https://commons.wikimedia.org/wiki/File:Freeman_Dyson_at_Harvard.jpg)

first task the problem of calculating the Lamb shift in a relativistic toy model with spinless electrons.

Nothing really useful came out of this, but it was Dyson's first close contact with quantum electrodynamics. He soon became interested in the methods used by Feynman, who was five years his senior, and began to get to grips with QED.

Feynman and Dyson had a lot in common. Both were brilliant scientists, and at the same time freethinkers with a keen interest in a wide variety of topics. For Dyson, this also included mathematics. In addition, both were excellent at calculating and had a deep understanding of physics, with Dyson tending to focus more on the mathematical side, while Feynman thought more in pictures. Interestingly, although Dyson became a professor during the course of his life, he never formally obtained his doctorate – there was just no room for it on his curriculum vitae. However, this did not in any way diminish his impressive reputation in the scientific world, where his outstanding achievements spoke for themselves. Many even believe that Dyson also deserved the Nobel Prize, along with Feynman, Schwinger, and Tomonaga.

Dyson used his time at Cornell University for an intense exchange with Feynman. In the summer of 1948, Feynman invited him to come along on a trip to New Mexico, where Feynman was heading because of a relationship with a young woman, although in retrospect it did not come to much. However, the trip made a great deal of sense for Dyson, who had a four-day opportunity to learn about Feynman's intuitive method from the master himself and to experience how powerful it was when it came to solving practical problems.

During their journey he repeatedly got into intense discussions with Feynman. As a mathematician, it bothered him enormously that Feynman's method had no mathematical foundation: "You have to get the math right. It is not enough to be able to calculate the correct results", he said to him again and again. And since Feynman was not very enthusiastic about this idea, Dyson decided to try it himself: Feynman's method had to be translated into a language that the normal physicist could understand – it needed a solid mathematical foundation. Guesswork and intuition alone were not enough.

At a 1948 summer school in Ann Arbor, Michigan, Dyson had the opportunity to listen to Schwinger's lectures and talk to him. Even for the mathematically adept Dyson, it was not easy to understand Schwinger's baroque formalism, but Schwinger was very nice to Dyson and went to great lengths to explain the details to him. Step by step, Dyson worked his way through Schwinger's lectures and finally came to believe he had understood the key points.

Meanwhile, it had become known that, besides the two approaches due to Feynman and Schwinger, there was also a third. In the spring of 1948, immediately after the Pocono conference, Oppenheimer had received a letter from a Japanese physicist called Shin'ichirō Tomonaga (Fig. 3.22) informing him of Tomonaga's QED research. Tomonaga had already written an important paper in 1943 and published it in Japan during the Second World War – much earlier than Schwinger. Meanwhile, the paper had been translated into English and had appeared in *Progress of Theoretical Physics* in the winter of 1946, but so far no one had taken any notice of it. Only when Oppenheimer forwarded Tomonaga's letter to all the participants of the Pocono Conference did this situation begin to change.

Like Schwinger, Tomonaga used a relativistic formulation of quantum electrodynamics based on the mathematical methods of quantum field theory. But Tomonaga's work – unlike Schwinger's – was written in a clear language that other physicists could easily understand, and he achieved the same results as Schwinger and Feynman. This is all the more remarkable in

Fig. 3.22 Shin'ichirō Tomonaga in 1965 (© picture alliance/Mary Evans Picture Library)

that Japan was largely isolated from the rest of the world during the Second World War. Apparently, despite this isolation, great progress had been made there in developing QED.

Three Equivalent Languages

So Schwinger, Feynman, and Tomonaga had told the world three times that QED works, only in a different language each time. There had to be a way to translate between these formulations, and Dyson was determined to find it. No one else in the world knew all three methods as well as he did. In September 1948, during a bus journey from San Francisco to Chicago that lasted several days, he had plenty of time to think about it, and suddenly everything came together in his mind to form a consistent structure. All three methods could be combined into a single theory once they were translated into a common language. It was precisely the language Feynman had avoided: the mathematics of quantum field theory. Dyson was very familiar with this, and so he succeeded in expressing all three approaches in this common mathematical language – some details can be found in Infobox 3.3.

Infobox 3.3: Feynman, Schwinger, and Tomonaga from Dyson's point of view

The language Dyson used to describe the approaches developed independently by Feynman, Schwinger, and Tomonaga was the mathematics of quantum field theory with its creation and annihilation operators for particles – exactly the formalism that Feynman had so long disliked. Dyson was very familiar with it and was able to classify the mathematical objects and terms used by Feynman, Schwinger, and Tomonaga using this language.

For instance, Schwinger's approach involved the so-called Green's functions, used in mathematics to solve certain differential equations, and Dyson recognized that these functions were closely related to the creation and annihilation operators for particles, thereby making contact with Tomonaga's approach.

Feynman, in turn, used propagators, i.e., elementary waves, which corresponded to the lines in his diagrams, where electron waves with negative energy run into the past. Dyson was able to associate these propagators with certain operator products, but they were built differently from those of Schwinger and Tomonaga. This made it possible to convert Feynman's formulation into the ones developed by Schwinger and Tomonaga, and vice versa.

For Dyson, the individual Feynman diagrams were nothing more than an illustration of the mathematical terms in an infinitely long perturbation series. For Feynman, on the other hand, they represented possible paths chosen by the particles, each with a certain quantum amplitude. In this way Feynman had come to his diagrams through vivid physical considerations, while for Dyson they inevitably arose from the mathematics of quantum field theory.

It was not easy to convince especially the older generation of physicists like Oppenheimer or Bohr that the problems of QED were now solved in principle if the procedure of renormalization was accepted. In a video interview, Dyson recalls a seminar in Princeton in which he tried to share his findings with Oppenheimer. But Oppenheimer refused to listen, vehemently contradicting him and hardly letting him have his say. It was an extremely frustrating experience for Dyson, and Oppenheimer's reputation dropped considerably in Dyson's eyes.

Like Niels Bohr or Werner Heisenberg, Oppenheimer was of the opinion that the existing problems with the infinities could not be solved within QED alone, and that a radically new approach was needed. This generation of physicists had initiated or at least witnessed the revolution brought about by quantum mechanics around 1925, i.e., a good twenty years earlier. Those were exciting times – many of them wanted to experience something like that again. And now a young fellow like Dyson came along and claimed that no revolution was needed, only a deeper understanding of how quantum theory worked within the framework of special relativity.

Dyson knew he was right, so he hurried to write down his thoughts in a clear and understandable way. In October 1948, he submitted his paper *The Radiation Theories of Tomonaga, Schwinger, and Feynman*, to *Physical Review*, where it appeared in February 1949. Thus Dyson presented Feynman's methods to the public even before Feynman had published them himself – Feynman's papers *The Theory of Positrons* and *Space-Time Approach to Quantum Electrodynamics* were not published until September 1949, about six months later.

Now it finally became clear that Feynman's methods also had a solid foundation and were mathematically sound – a real breakthrough for Feynman. His methods were as trustworthy as those of Schwinger and Tomonaga, because they were all just different formulations of the same theory. Feynman's method had the great advantage that it was easy to use – one only had to draw all relevant diagrams, translate them into formulas according to well-defined rules, and then carry out the calculations. Nowadays, computers do this tedious work, making it possible to handle huge numbers of Feynman diagrams, something that would probably have made Feynman himself dizzy. Later on, Schwinger commented on this, possibly with a slight touch of jealousy:

> Like the silicon chips of more recent years, the Feynman diagram was bringing computation to the masses.

In February 1949, Dyson went one step further, proving in a second paper that the process of renormalization also works for more complex Feynman diagrams. The article was also published in *Physical Review* in June 1949, under the title *The S Matrix in Quantum Electrodynamics*. Everything about electrons and photons could thus be very accurately calculated using QED, if enough Feynman diagrams were taken into account. To this day, QED is one of the very best theories that physics has ever produced. The only thing is that the mass and charge of the electron must be used as input to carry out the renormalization – these parameters cannot be calculated within QED.

When a follow-up conference to Shelter Island and Pocono was held at Oldston-on-the-Hudson in Peekskill, New York, in April 1949, it became a triumph for Feynman and Dyson. It was only two years previously that the groundbreaking experiments of Rabi and Lamb had given the decisive impetus at the Shelter Island Conference. In the previous year at the Pocono conference, Julian Schwinger had attracted all the attention, while Feynman still stood in his shadow and Dyson was not even invited because of his youth. But now at the Oldstone Conference – not least thanks to Dyson – Feynman's great hour had come. His methods have since established themselves, not only in QED, but throughout quantum field theory, and there is scarcely a publication in this area that does not contain at least one Feynman diagram.

4
California, Super-Cold Helium, and the Weak Interaction

Feynman's masterpiece in QED was now accomplished and he was attracted to new, warmer climates – first to Brazil, then to Caltech (the California Institute of Technology) in California, where he would stay for the rest of his life. Feynman married his second wife Mary Louise Bell, but the marriage turned out to be unhappy and lasted only four years.

Feynman also reorientated his work in physics and became enthusiastic about various topics. He would go on to achieve two further groundbreaking successes: with the help of his path integrals he managed to explain the strange behavior of extremely cold helium on the basis of quantum mechanics, and in particle decays caused by the weak interaction, he and Murray Gell-Mann showed how nature makes a big difference between right and left.

4.1 Brazil and the Move to Caltech

By the end of 1949, Feynman had largely completed his contribution to QED. Dyson had already provided the mathematical foundation, but Feynman could not resist deriving his method again in his own words and publishing it in *Physical Review* in November 1950 under the title *Mathematical Formulation of the Quantum Theory of Electromagnetic Interaction*. About a year later, he published the paper *An Operator Calculus Having Applications in Quantum Electrodynamics*, which dealt with more formal aspects of the theory and an introduced an elegant new

J. Resag, *Feynman and His Physics*, Springer Biographies,
https://doi.org/10.1007/978-3-319-96836-0_4

notation – throughout his life, Feynman always liked to invent his own notations when the usual ways of writing things down seemed too elaborate to him.

In early 1949, Feynman began to yearn for change. Although he was already 31 years old, he had never left the USA, so he gratefully accepted the invitation of the Brazilian physicist Jaime Tiomno to spend the summer weeks as a visiting scientist in Rio de Janeiro. With its warm climate and relaxed lifestyle, Brazil was exactly to Feynman's taste, and he enjoyed his time there, giving physics lectures in the morning and relaxing on the beach in the afternoon.

Back in Ithaca he was filled with a certain restlessness. Autumn and winter came, cold and frost covered the landscape, and Feynman missed the sunshine and warmth of Brazil. When he was surprised by a snowstorm in his car and tried to put on the snow chains, he reached the point where he'd had enough. It was freezing cold, his fingers became stiff, and the stupid snow chains simply would not be attached to the wheels. There had to be a place in the world where he wouldn't have this problem![1]

In addition, Feynman was unhappy with certain things at Cornell University. There were all kinds of departments that didn't much interest him, teaching such subjects as domestic science, philosophy, music, and so on. Feynman never had much to do with the people working in these areas.

Cornell or Caltech?

In the winter of 1950, Feynman was invited by Robert Bacher, whom he already knew from Los Alamos, to give a few lectures at Caltech (California Institute of Technology) in Pasadena near Los Angeles. As a rather small but top class university, Caltech was specialized in natural sciences and engineering, something that suited Feynman down to the ground. Nobody was interested in domestic science at Caltech!

When Feynman arrived at Caltech, Bacher lent him his car so that he could spend his evenings in the bars and nightclubs of the Sunset Strip. This was exactly what Feynman liked best, and he soon began to think about moving to sunny California – an idea that Bacher strongly supported. Before long, Feynman received an offer to do just that.

[1] See *Surely you're joking* in the chapter *An Offer You Must Refuse.*

When they heard about this at Cornell, they made him offers to stay there as well. Feynman felt like the donkey standing exactly in the middle of two piles of hay, unable to decide which one to take. Whenever he had just made up his mind, the other side would go one better.

California was tempting with its pleasant climate, but he still liked Cornell too – he would miss Hans Bethe in particular. Moreover, in the meantime, he had earned the right to take sabbatical leave, and he had plans to spend this in Brazil once again. That was the deciding factor and Feynman thus chose to stay in Ithaca.

But it wasn't that easy for him. When Caltech heard of his decision, they wrote to say that they would hire him immediately and give him his first year as a sabbatical year, so that he could spend a year in Brazil. This was indeed unbeatable, and Feynman accepted the offer.

In the fall of 1950 he moved to Caltech and spent the months from August 1951 to June 1952 in Rio, where he stayed in a hotel in Copacabana, overlooking the beautiful beach, spending the evenings in the bars and making many contacts with beautiful young women. He learned Portuguese and took the opportunity to perfect his bongo playing in the land of the samba. Rhythm was in his blood! He eventually became a really good bongo player – you can get an impression of this on YouTube (look for *Feynman PLAYS THE BONGOS*).

Feynman enjoyed life to the full – or at least, so he tried, because in retrospect some things may look rather like an attempt to compensate for the tragic death of his wife Arline and to overcome his subsequent depression. It seems also that he overdid things with alcohol. Walking along Copacabana beach one afternoon, he saw a bar and suddenly felt a strong desire for a drink. That would be just the right thing, he thought. But it was only the middle of the afternoon, and there was nobody there but him. That was when he began to get scared. How come he had such a terribly strong feeling that he just had to have a drink?[2]

Since that experience, Feynman never drank any alcohol again. He was afraid that it could ruin his mental abilities: "You see, I get such fun out of thinking that I don't want to destroy this most pleasant machine that makes life such a big kick."

[2]See the chapter *O Americano, Outra Vez!* in *Surely you're joking*.

Mary Lou

The many female acquaintances and affairs which he had begun over the previous few years did not bring him what he was looking for. There was simply no one who could compete with Arline. It seems that the letter he had written to the deceased Arline in October 1946 still described how things looked in him:

> I don't understand it, for I have met many girls and very nice ones and I don't want to remain alone – but in two or three meetings they all seem ashes. You only are left to me. You are real.

But now Feynman finally wanted to escape from this trap, which was seriously affecting his life, and there was a kind of short-circuit reaction: towards the end of his stay in Brazil he remembered a very attractive young woman whom he had met in Ithaca and who, by pure chance, had also moved to California. Her name was Mary Louise Bell, but he usually just called her Mary Lou.

Feynman had been attracted to her for some time, but there had been repeated arguments and they had finally separated. Maybe they shouldn't have argued so much, Feynman thought from his distant vantage point in Brazil. Actually, she was a wonderful woman, and very attractive, and she wasn't stupid either: she had studied art history and Feynman had learned a lot about art from her. So he wrote her a letter and asked her to marry him.

In June 1952, they married – almost exactly ten years after his marriage to Arline. But their marriage was not happy, because Mary Lou had a completely different approach to life than Richard. It was important for her to be the wife of a respectable professor, who had to behave in a certain way and follow a certain dress code, the very things Feynman did not like at all – as we know, he had always hated formalities since childhood. Moreover, Mary Lou was completely unfamiliar with the scientific community and did nothing to support Feynman in his contacts with his colleagues – for example, she called the famous Niels Bohr the "old bore". There was no way this could last for long. After two years they had already separated, and they finally divorced after just four years, in the summer of 1956. Feynman must have felt a great sense of relief.

The attempt to settle in California and build up a regular family life with Mary Lou had therefore failed. Not surprisingly, Feynman frequently doubted whether the move from Ithaca to California had been the right decision. Should he perhaps return to Ithaca and make good his mistake? There were no such cold winters in California as in Ithaca, but in the summer heat a heavy smog developed, which made breathing difficult and burned the eyes, so California was not perfect either!

On a particularly bad day for the smog, Feynman actually called Cornell University and asked if he could come back. "Sure!" they said, and began to get excited. "We'll set it up and call you back tomorrow!" But before it came to that, Feynman had several key experiences that swept away that whole idea: "God must have set it up to help me decide."

Caltech – Here I Want to Be

So what happened? When Feynman went to his office at Caltech the day after the call, he met someone on the campus who told him about an amazing discovery: the German physicist and astronomer Walter Baade had found that there were two different types of a certain kind of variable stars – the so-called Cepheid variables – and not just one as had previously been thought. This was important for the following reason. The brightness of these giant stars pulsates with periods from days to weeks, in such a way that, the larger and therefore brighter they are, the more slowly they pulsate. And this meant that the absolute luminosity of these stars could be deduced from their pulsation rate. From their visible brightness in the sky, or apparent luminosity, it was then possible to determine how far away they were. Since Cepheid variables are a thousand times brighter than our own star, the Sun, they can also be seen in many other galaxies and thus can be used to calculate the distances to these galaxies.

The problem was that the distances came out too small. If one determined the expansion rate of the universe with the help of these distances and extrapolated back to see how old the universe had to be, one only got out about three billion years. However, it was already known that the Earth was about 4.5 billion years old. That didn't fit – how could the universe be younger than the Earth? Here was a major problem!

The solution was that there are two different types of Cepheid variables, as Baade had discovered. Type I Cepheids are significantly brighter than type II Cepheids at the same pulsation frequency, due to the different heavy element content in these stars (Fig. 4.1). This was what Baade had shown. By considering the Cepheid type correctly, it turned out that the galaxies were

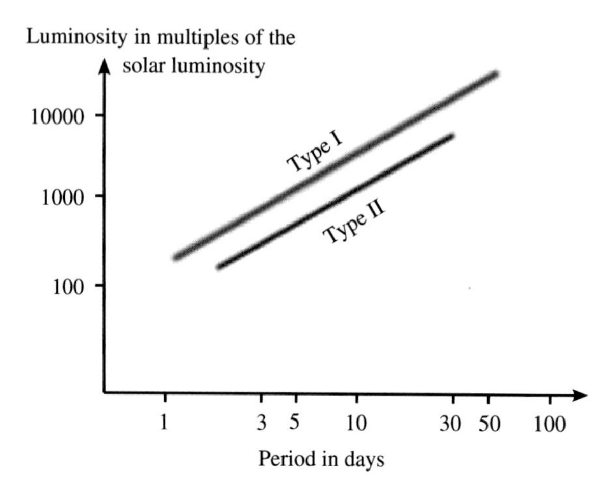

Fig. 4.1 Relationship between pulsation period and luminosity in type I and II Cepheids

at least twice as far away and that the universe was therefore significantly older than the Earth.

This was fantastic! Feynman had just learned on the way to his office that the universe was clearly older than previously assumed, and that there was no longer any contradiction with the age of the Earth.

But then, a few steps further, came yet another high point. The young biologist Matt Meselson came up and told him what he had just been able to demonstrate in a sophisticated experiment. When a bacterium divides, it does not simply produce a complete copy of its double-strand DNA. Instead, the double-strand is separated into two single strands and the missing halves are then completed, resulting in a total of two double-strands. Each daughter cell receives one of these two double-strands. At the time this was a fundamental discovery, and Feynman was one of the first to hear about it on the Caltech campus!

When Feynman (Fig. 4.2) finally arrived in his office, he knew that this was exactly where he wanted to be – the place of work of outstanding scientists from various disciplines who would be able to keep him informed about all kinds of exciting discoveries. When Cornell called later, he had to confess that he had changed his mind. And that was the way it had to be. Never again would anything change his decision to remain at Caltech, he swore to himself. And indeed, Feynman did stay with Caltech for the rest of his life, resisting all attempts by other universities to entice him away, sometimes with exceptionally high salary offers. And it was the right decision!

Fig. 4.2 Richard Feynman in 1954. On the blackboard are typical integrals of the kind that occur when calculating Feynman diagrams (© Estate of Francis Bello/Science Photo Library)

As far as physics was concerned, Feynman could have taken it easy after 1950. He had already achieved remarkable successes – more than most physicists achieve in their whole lifetime. His position at the university was guaranteed and he was a widely respected scientist. In addition, Feynman had now passed the age of thirty – a barrier in terms of scientific performance. The situation is very similar in high-level sports: from a certain age you may have greater experience, but your mind and body are not quite as powerful as they used to be. This makes it difficult to remain at the forefront of research for any length of time. Many of the great discoveries were made by physicists before their thirtieth birthday: Dirac was 26 when he established the Dirac equation, and so was Heisenberg when he formulated his quantum mechanical uncertainty principle. And how old was Albert Einstein when he developed his special theory of relativity? You've no doubt guessed it: he was 26.

Now Albert Einstein stands as an example that something significant can still be achieved by someone over the age of 30: at the age of 36, he published his general theory of relativity, in which he explained gravity by the curvature of space and time, and at the age of 45, together with the Indian physicist Satyendranath Bose, he predicted a quantum mechanical state of matter that could only exist at extremely low temperatures, viz., the Bose–Einstein condensate, to which we shall return soon.

Albert Einstein was certainly an exception, still making outstanding contributions in the middle of his life. But Feynman was also far from considering retirement at thirty – he loved physics too much for that. The problems of QED were largely solved, so what problem should he turn to next?

The Particle Zoo Keeps on Growing

One possible field of research was provided by the newly discovered mesons. These short-lived particles are known today to consist of a quark and an antiquark, but that was not known at the time. However, it was known that these particles have an integer spin and can be created when protons collide with other protons or with atomic nuclei. According to Einstein's formula $E = m \cdot c^2$, they materialize from the available collision energy as soon as this is sufficient, and decay again within a fraction of a second. In nature, such collisions occur in the upper atmosphere when high-energy protons from space hit the atomic nuclei of nitrogen and oxygen (Fig. 4.3). The first mesons were thus discovered – just like the positron – in cosmic radiation.

In the meantime, scientists had reached a point where they were no longer solely dependent on these natural collision events. The new particle accelerators had progressed considerably in recent years, generating collision energies sufficient to produce whole hosts of different mesons. Gradually, an entire zoo of new particles appeared in the experiments, including lots of other particles as well as mesons.

Indeed, the variety of particles became somewhat confusing. The situation had been much easier before. We knew about the proton and the neutron in the atomic nucleus, the electron in the atomic shells, and the photon as the particle of light. Around 1930 the neutrino came along, introduced to explain the energy distribution in radioactive beta decay, although this ghostly particle was only actually detected some 26 years later. There were also the associated antiparticles, like the positron. All these particles were considered elementary at the time. They were thought to make up the whole world.

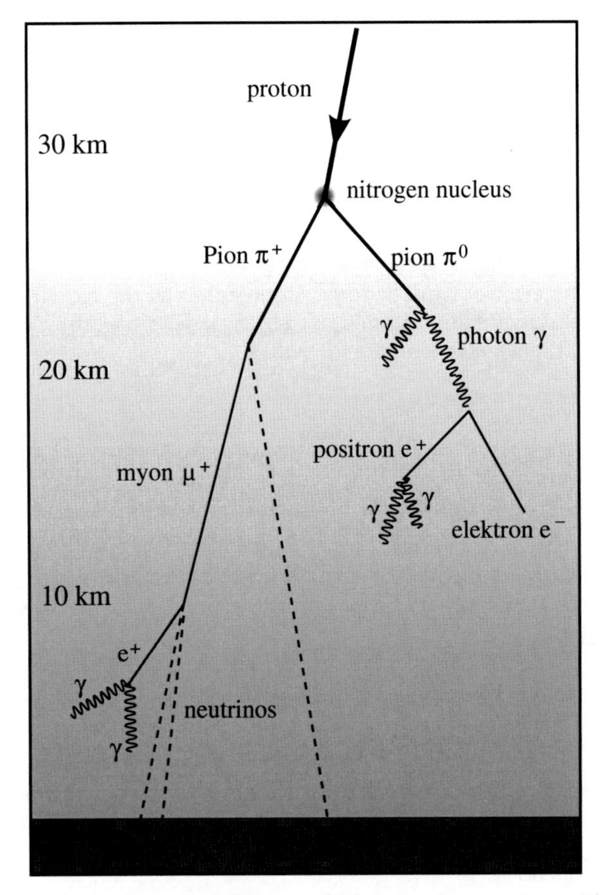

Fig. 4.3 In cosmic radiation, the collision of high-energy protons with molecules in the air creates cascades of unstable particles. These include pions, a kind of meson

The Japanese physicist Hideki Yukawa – a close friend of Tomonaga – had already speculated in 1935 that there could be at least one other particle. Yukawa had been wondering how the strong nuclear forces were transmitted between protons and neutrons in the atomic nucleus. Could there be an analogy with electric forces, which work by the exchange of massless photons (see Fig. 4.4)? What properties would a similar exchange particle have to have for the strong nuclear force?

The main difference with the electromagnetic force is that the strong nuclear force has only a very short range. It drops to almost zero at distances of only a few proton radii. So we have to get very close to an atomic nucleus to feel this nuclear force. Apparently then, the virtual exchange particle can only cover very short distances – it runs out of breath over longer distances.

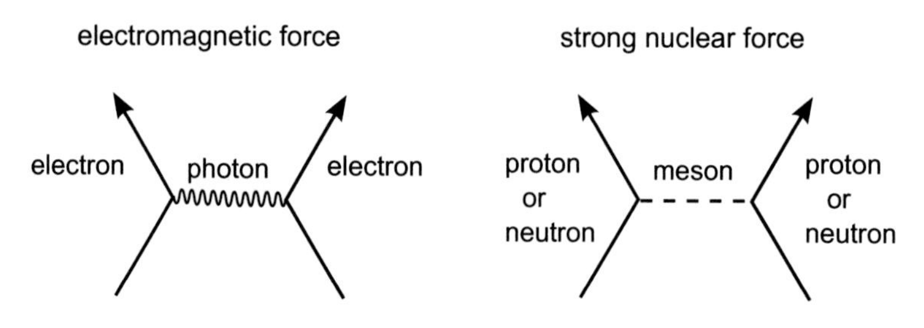

Fig. 4.4 Just as the electromagnetic force is generated by the exchange of photons, the strong nuclear force in the atomic nucleus was generated by the exchange of mesons (especially pions), according to Yukawa

This is where the quantum mechanical energy–time uncertainty comes into play. The more energy the exchange particle has to borrow for its short existence, the shorter this existence has to be, and the smaller the distance it can cover during this time. With the photon this is not a problem, because it is massless. However, if the exchange particle has a mass, it must borrow the energy stored in this mass.

With this reasoning, Yukawa could deduce the mass of the hypothetical exchange particle from the known range of the strong nuclear force: it should lie in the range 100–200 MeV (expressed in the energy unit mega electronvolt, i.e., million electronvolt). Thus the mass would be between the electron mass (0.511 MeV) and the proton mass (938 MeV), and this is exactly what the word *meson* is supposed to express, from the Greek 'to méson' or 'mean'.

From today's perspective, Yukawa's explanation of the strong nuclear force in terms of meson exchange was only an intermediate step. The attractive force is now known to take place between the building blocks of protons and neutrons – the quarks – and is mediated by gluons. In the nucleus, the attractive effect of the gluons also reaches the quarks of directly adjacent protons and neutrons.

In 1935, however, nothing was yet known about either quarks or gluons, so Yukawa's model of the nuclear force was the best available. In 1936, in the search for the hypothetical exchange particles of this model only one year after Yukawa's prediction, a particle with a mass of 106 MeV was indeed found in cosmic radiation. This is called the muon today. In time, however, it turned out that the muon did not have the right properties to be the virtual exchange particle that generates the nuclear force. It behaved rather like a heavy version of the electron, and was not involved in any way in the nuclear force.

So the muon was actually a complete stranger and quite unexpected. As quipped by Isidor Rabi, whom we already know from the *g*-factor and Lamb shift measurements: "Who ordered that?" Only a good ten years later (in 1947) were Yukawa's particles finally found in the form of the three pions π^+, π^-, und π^0, which have masses around 140 MeV, and are thus about seven times lighter than the proton.

Since then, new accelerators have been producing one new particle after another for years. The whole situation was quite confusing. Could all these particles be elementary entities? How did they fit with the other particles? Did they have anything in common?

Many physicists rushed into these questions, and Feynman was also interested. Unfortunately, the situation was quite chaotic: new experimental data were continuously being published, and it was somehow difficult to recognize any pattern in them. Feynman began to get the impression that the time was not yet ripe for a comprehensive theory that could bring order to this chaos. Besides, there were just too many people in the same boat, and that put him off – he preferred to pursue his own ideas and didn't particularly like reading his colleagues' publications.

So Feynman set out in search of an area that was not so much the focus of general interest, and yet still offered interesting problems. And this he found when he started working on the physics of very low temperatures, close to absolute zero. Here there were amazing macroscopic quantum phenomena just waiting for a theoretical explanation. With his extensive experience in quantum theory, Feynman felt himself well equipped for this, and was eager to explore the reasons for these phenomena in his typical unconventional Feynman way.

4.2 Physics at Low Temperatures

What happens if we keep cooling a gas? The answer seems simple: the average kinetic energy of the atoms or molecules continues to decrease until the attractive forces between them cause them to condense into a liquid. If the system is further cooled, these components will eventually be forced into a lattice structure and the liquid will freeze to form a solid. Sometimes the liquid state can also be skipped – the details depend on the substance under consideration and the external pressure, and they will be of no further interest to us here.

A solid can be cooled still further, but there is a lower limit, the absolute zero temperature, which is zero kelvin (minus 273.15 °C). According to classical ideas, which would later be modified by quantum mechanics, the atoms have completely lost all kinetic energy at absolute zero. They sit motionless at their permanent positions within the crystal lattice, which becomes completely rigid. Any activity, any molecular vibration, rotation, or vibration, comes to a complete standstill. Further cooling is no longer possible.

Exactly when the transitions to the liquid and then to the solid actually take place depends on the attractive forces and the chemical bonds between the atoms. At normal atmospheric pressure, carbon is only gaseous above a glowing hot 3915 K (3642 °C), while nitrogen must be cooled to an icy 77 K (minus 196 °C) to get it to condense to a colorless liquid, and finally freezes to a solid only below 63 K (minus 210 °C).

So is that all that can be said about physics at low temperatures? That at some point every substance will eventually solidify if we just cool it down far enough? That sounds perfectly credible, but as so often, nature surprises us when we enter new areas that were previously inaccessible.

One of the first to venture into the area of these lowest temperatures was the Dutch physicist Heike Kamerlingh Onnes. At the beginning of the twentieth century, he set up a laboratory for cryotechnology at Leiden University, and it soon became the world leader in liquefying gases and exploring low temperatures.

In 1908 he succeeded in liquefying helium there for the first time. This required extremely low temperatures, since at normal atmospheric pressure helium only becomes liquid at 4.2 K (minus 269 °C). This temperature is significantly below the temperature of 77 K at which nitrogen is transformed into a liquid, and it is no easy matter to reach such low temperatures in the laboratory.

The reason for the very low boiling point of helium is essentially that helium is a noble gas – and the lightest of all noble gases. After hydrogen, helium atoms are even the lightest of all atoms. The temperature must therefore be very low for these light helium atoms to settle down, so that the weak attractive forces between them can unite them into a liquid. This makes helium the record holder – no other substance remains gaseous for as long as helium when the temperature drops, not even hydrogen (see Fig. 4.5).

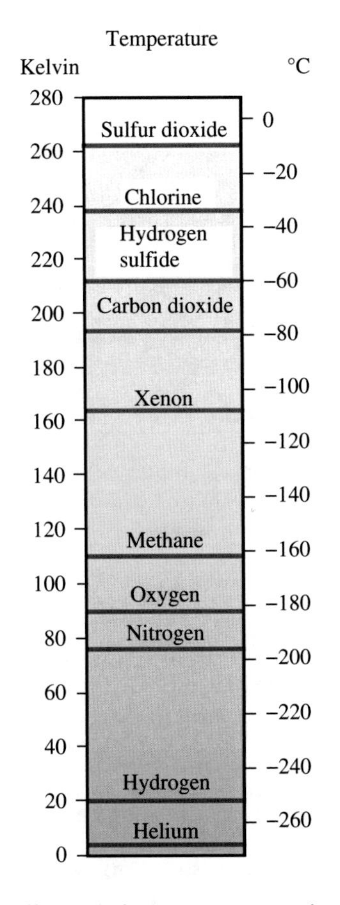

Fig. 4.5 Boiling points of different substances at normal pressure

Supercold Surprises: Superconductivity and Superfluidity

Heike Kamerlingh Onnes had thus cracked the hardest nut that could be cracked when it came to liquefying gases. And he could now use liquid helium to cool other materials down to very low temperatures. In 1911, when Onnes did this with mercury to investigated its properties, he made an astonishing discovery. Just below 4.2 K, the metal suddenly lost all electrical resistance. Once an electric current had been set up, it could go on indefinitely with no power source, because it didn't lose any energy. Onnes had discovered superconductivity!

Nobody had expected anything like this. It was known that the electrical resistance of metals normally decreases with decreasing temperature, because

Fig. 4.6 When the critical temperature falls below 2.2 K, liquid helium transforms from a wildly bubbling liquid (left) into a completely calm one (right). Source: *Credit*: Alfred Leitner, left: https://commons.wikimedia.org/wiki/File:Liquid_helium_lambda_point_transition.jpg, Bmatulis on Wikipedia, right; https://commons.wikimedia.org/wiki/File:Liquid_helium_superfluid_phase.jpg

the diminishing vibrations of the metal atoms cause less and less disturbance to the flow of electrons. However, the fact that the electrical resistance could drop to exactly zero below a certain temperature was a completely new feature.

So there was definitely something new to explain when matter was cooled to very low temperatures. And superconductivity would not remain the only unexpected phenomenon. Helium itself would turn out to be full of surprises.

In the 1920s and early 1930s, it gradually became clear that liquid helium had some strange properties. These show up when it is cooled significantly below its boiling point of 4.2 K. To achieve this, we simply connect a good vacuum pump to the thermally insulated vessel containing the liquid helium and suck off some helium gas. The boiling point of the helium sinks due to the falling pressure, and the continuously evaporating helium cools down the remaining liquid helium. As soon as the temperature approaches the value of about 2.2 K, something strange happens.

The liquid helium "boils" more and more because, despite all the thermal insulation, some heat always gets in from the outside and causes helium gas bubbles to form on its walls, as happens at the bottom of a pan of water on a hot plate. But then, when the temperature falls below 2.2 K, this boiling suddenly ceases. The liquid, which previously looked like boiling water, suddenly becomes very calm, even though some heat will still be getting through the walls (Fig. 4.6).[3]

[3]You can watch this phenomenon in impressive videos on the Internet, for example at http://www.alfredleitner.com/superfluid.html.

What's going on here? More detailed investigations showed that the properties of liquid helium change abruptly when the temperature falls below the critical level of 2.2 K. A distinction is therefore made between helium I – liquid helium above the critical temperature – and helium II at temperatures below. While helium I is still a normal liquid, the properties of helium II are quite exotic.

A case in point is its thermal conductivity. Helium II can conduct heat extremely well, far better than copper, for example. This is also the reason why the formation of bubbles stops so suddenly: the heat penetrating through the vessel walls is immediately dissipated and distributed through the liquid, so that it is no longer warm enough on the walls to form helium gas bubbles. Only at the surface of the liquid does helium continue to evaporate and thus dissipate the penetrating heat.

The excellent heat conductivity of helium II is very unusual for a liquid, because liquids don't normally conduct heat very well. But this is different in helium II. Heat is conducted in a similar way to the much faster propagation of waves, like sound waves, and we also speak of the *second sound.* For example, real heat pulses can be sent through the liquid, propagating at speeds of several meters per second.

And there are other unusual properties. When a fluid flows through a tube, its internal friction – called viscosity – usually slows it down, and the narrower the tube, the more it slows down. A liquid cannot normally penetrate very fine pores with diameters of less than 1 μm. If a liquid is poured into a vessel whose base is traversed by such fine pores, this base will normally be impenetrable to the liquid.

This is also the case with helium I – it cannot penetrate such fine pores and remains in the vessel. But as soon as the helium cools below 2.2 K and turns into helium II, something astonishing happens: the porous base suddenly becomes leaky. The liquid helium can penetrate the fine pores and it drips down as if the pores no longer represented any kind of barrier. This remains the case when the pores are made even finer. The liquid helium seems to have completely lost its viscosity – in a sense, it has become more liquid than liquid, which is why this exotic state of matter is referred to as *superfluidity.*

The loss of internal friction can also be seen in another effect. If superfluid helium II is poured into a vessel that is open at the top, a very thin film of liquid will form on the vessel wall. At a thickness of around 30 nm, the film is so thin that it is practically invisible to the eye. It is caused by the tiny forces of attraction between the helium atoms and the atoms of the vessel wall.

Fig. 4.7 Superfluid helium creeps over the edge of the vessel and drips down. *Credit* Alfred Leitner; https://commons.wikimedia.org/wiki/File:Liquid_helium_Rollin_film.jpg

Since superfluid helium has no frictional resistance, the liquid film is pulled further and further upwards. Finally it reaches the edge and also covers the outside of the vessel. Gradually more and more helium is drawn up over the edge of the vessel, whereupon it slides down the outside and collects at the bottom of the vessel in a droplet that finally detaches (Fig. 4.7). If we wait long enough, all the helium will crawl out of the vessel and drip down.

It is therefore quite difficult to confine superfluid helium within a vessel, since it can penetrate even the finest pores and it can also creep out over the vessel walls and escape into the open. In both processes it shows no viscosity.

However, it would be a simplification to say in general that helium II has no viscosity at all. For example, if a metal cylinder is set in rotation in helium II, it seems that after some time the liquid is drawn along with it and gradually set in rotation. This can be seen, for example, by the fact that the co-rotating liquid is able to set a small propeller wheel in rotation. If helium II had no internal friction at all, this would not be possible – the rotating metal cylinder would have no way of setting the surrounding liquid in motion.

How can these contradictory results about viscosity be explained? In order to solve this problem, the Hungarian-born American physicist Laszlo Tisza came up with the following idea in 1938. Helium II could in some sense be a mixture of two liquids. One of them is superfluid and can creep through the finest pores and glide up the walls of vessels without any friction. The other component, on the other hand, has the properties of a normal liquid and has a certain – albeit low – viscosity.

This simple model can actually explain many of the properties of helium II. It is found that the superfluid component behaves like a medium without any thermal energy. The helium atoms in this component do not appear to perform any random motions or oscillations to which a thermal energy and temperature could be assigned. If they move at all, they clearly do so only in a strictly ordered manner, hence in step, as it were.

The normal component, on the other hand, behaves like a gas whose particles reside in the suprafluid component and can move independently through it. Therefore, this component is perfectly capable of storing and transporting thermal energy.

In this model, the extremely effective heat conduction of helium II can be explained like this. At the heat source, the normal component absorbs thermal energy and moves away from the source to colder areas to distribute the heat. In order to keep the total helium density constant, the superfluid component flows in turn to the heat source, where it is partly converted into the normal component by the heat. Thus, there are two opposing flows of normal and superfluid components, and their velocity is limited only by the low viscosity of the normal component – the heat energy of the source is therefore dissipated and dispersed very quickly.

When all these phenomena are analyzed in detail, it is found that the closer the temperature comes to absolute zero, the greater the proportion of the superfluid component will be (Fig. 4.8). However, this picture should not be overinterpreted either. There are not really two different liquids in helium II that could somehow be separated from each other. It is basically a simple model that describes the behavior of helium II quite well in many cases.

Quantum Mechanics Is Essential

Is there a way to establish the physical foundations of this model? Why do more and more helium atoms move in step the colder it is? And why does part of the helium not seem to do this, so that we find a normal component coexisting with the superfluid component? These were the questions that Feynman found himself asking at the time. One thing was clear to him: the

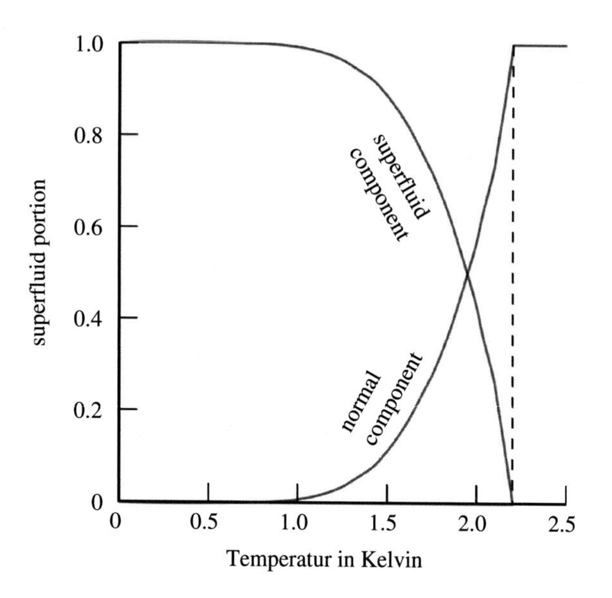

Fig. 4.8 According to the two-fluid model, below 2.2 K (dotted line), a superfluid component is formed in liquid helium, the proportion of which increases upon further cooling

exotic behavior of liquid helium could not be explained on the basis of classical physics. Quantum mechanics had to play a crucial role.

A first clue that quantum effects are involved is provided by another remarkable phenomenon: helium is the only substance that does not freeze at normal pressure, even if we cool it down to zero kelvin. Only under an additional pressure of at least 2.5 MPa (25 bar) is it possible to convert helium into a solid.

Classically, such behavior is impossible. At zero kelvin, any thermal motion of the helium atoms should be reduced to zero, so that even the relatively weak attractive forces between them should be sufficient to combine them into a crystal. Since this does not happen, only one conclusion remains: classical physics is not sufficient to understand the phenomenon. We must include quantum mechanics.

In quantum theory, there are no absolutely stationary localized objects. This is a consequence of Heisenberg's uncertainty principle, which says the following: the more precisely we determine the location of a particle, the less precisely its momentum is defined – and vice versa (for details, see Infobox 4.1, and see also the section on energy–time uncertainty and virtual particles in Chap. 3).

Infobox 4.1: Heisenberg's uncertainty principle

Heisenberg's uncertainty principle was formulated by Werner Heisenberg in 1927. It says that the product of the uncertainty Δx in the location and the uncertainty Δp in the momentum of a particle is approximately equal to the Planck constant h:

$$\Delta x \cdot \Delta p \approx h.$$

Ultimately, this principle is a consequence of a universal property of waves, according to which a spatially limited wave packet comprises a broader frequency spectrum the smaller it is. With sound waves, for example, a loud bang contains quite a broad spectrum of different frequencies – and all the more, the shorter it is. In quantum theory, each frequency or wavelength of a quantum wave now belongs to a certain momentum of the associated particle, as Louis de Broglie pointed out in 1924. A shorter wavelength λ belongs to a higher particle momentum p. The exact correlation is

$$\lambda = h/p.$$

A wave packet therefore contains more wavelengths and thus more particle momenta the more restricted it is in space. It is not possible to tell exactly where the particle is located within the wave packet, nor what momentum it has. Rather, the wave packet indicates the possible locations that may be found in the corresponding position measurement, while its momentum spectrum indicates the possible results of a momentum measurement. But besides that, the exact values found for the particle's location or momentum are absolutely random – this is exactly what is meant when we talk about the random quantum mechanical motion of a particle in the main text.

This can be imagined in the following way: the smaller the region in which a particle is required to remain, the more it will begin to move uncontrollably. It will vibrate and jump back and forth if we try to nail it to one place. This is exactly why the electrons of the atomic shells do not fall into the atomic nuclei: the more closely an atomic nucleus draws an electron towards it, the greater will be its uncontrolled random motion and hence also its average kinetic energy. It will thus resist falling into the core. Ultimately, this results in a compromise between the kinetic and the potential energy, and this determines the size of the region in which the electron may reside within the atom.

So what does the uncertainty relation mean for the behavior of atoms in a liquid? As the temperature drops, the forces of attraction between the atoms try to join them together to form a crystal lattice, thus severely restricting the location of each atom. The atoms fight against this by their quantum mechanical random motion. Normally, however, the forces of attraction will be strong enough to prevail against this random motion and force the atoms into fixed positions in a crystal lattice.

Not so with helium atoms! Only very weak attractive forces act between them, and due to their low mass, the quantum mechanical random motion is quite intense. They tremble and wriggle so violently that, even at zero kelvin, such weak forces are not sufficient to keep the helium atoms at their fixed positions in the lattice. At normal pressure, helium therefore remains liquid even at absolute zero temperature.

Atoms in Equal Quantum Step: The Bose–Einstein Condensate

So the quantum mechanical uncertainty principle ensures that helium remains liquid even at absolute zero. Feynman was therefore on the right track when he sought to justify the behavior of liquid helium by means of quantum mechanics. But what about the other properties of liquid helium? Could quantum mechanics also help to understand why helium becomes superfluid below 2.2 K and could it explain what the normal and superfluid components are all about?

To answer these questions, let's go back briefly to 1924, when quantum mechanics was still in its infancy. In his famous dissertation, the young French physicist Louis de Broglie had just made the bold suggestion that light was not alone in having a dual nature as wave and particle. Indeed, a wave should also be assigned to particles like the electron, and the same relationship $\lambda = h/p$ should hold between the wavelength and particle momentum as for photons (see Infobox 4.1).

Many physicists were initially very sceptical about de Broglie's idea. It seemed to the older generation in particular that younger physicists like de Broglie were becoming rather too careless by making such far-fetched proposals and calling classical physics into question in this way. Not so the then already 45-year-old Albert Einstein, who wrote in 1924[4]:

> De Broglie's work made a considerable impression on me. He has lifted a corner of the great veil.

[4]"Die Arbeit von de Broglie hat großen Eindruck auf mich gemacht. Er hat einen Zipfel des großen Vorhangs gelüftet." See Einstein's letter about de Broglie's doctoral thesis, dated December 16, 1924 and addressed to the French physicist Paul Langevin, quoted in Henning Sievers: *Louis de Broglie und die Quantenmechanik*, https://arxiv.org/pdf/physics/9807012.

In the same year, 1924, the young Indian physicist Satyendranath Bose sent Einstein a paper he had written. Bose had used de Broglie's relationship between waves and particles to derive *Planck's radiation law* in a new way, using the methods of statistical quantum mechanics. This opened a new door that would also prove useful for describing completely different systems. Let's take a closer look.

Planck's law describes how heat energy is distributed over the individual frequencies in heat radiation – for example, in a hot oven. A key feature of this law is that, the higher the temperature of the radiation, the more energy is allocated to the high-frequency waves. An example should make this clear. Glowing charcoal burns at a temperature of about 800 °C. The heat radiation emitted is mainly invisible infrared light, but it also contains a small part that our eyes perceive as reddish light. If we blow air onto the charcoal with a bellows, its combustion temperature rises and it glows with a bright yellowish color. As the temperature increases, the energy content of the higher frequencies in the heat radiation increases, and this is noticeable, for example, in the color of glowing bodies.

When he announced his radiation law in 1900, Max Planck had solved a long-debated physical problem theoretically and opened the door to the quantum world almost by chance. For he could only derive the law if he assumed that atoms were compelled to emit or absorb light in certain energy packets. At that time, however, Planck did not yet believe that light itself only existed in the form of individual light quanta, i.e., photons.

Bose, on the other hand, was able to found his work on the concept of the photon more than twenty years later because Albert Einstein had introduced it in 1905, even though the existence of photons remained controversial for many years. Accordingly, heat radiation also consists of photons whose momentum and energy correspond to different wavelengths and frequencies. One can think of heat radiation as a photon gas, whence the walls of a hot oven constantly emit and absorb photons. The photons do not interact with each other significantly, and this is why we speak of an ideal gas.

Bose now succeeded in calculating how the energy of the heat radiation is distributed statistically over the individual photon frequencies at a given temperature. The more energy a photon has, the higher the corresponding wave frequency and the shorter the wavelength. In this way, Bose was able to derive Planck's radiation law directly from quantum mechanics using statistical methods.

In doing so, Bose used a universal concept of quantum mechanics that is perfectly established today, but was still much frowned upon in those days:

the indistinguishability of identical particles. Since particles don't have definite trajectories in quantum theory, we can never say which photon we have just measured in heat radiation. This indistinguishability has a decisive influence on the energy distribution of photons.

In order to publish his work, Bose asked Einstein for help. Some time earlier, Bose had translated Einstein's work on general relativity into English, with his permission. Now he asked Einstein to translate his own work into German and publish it, if he felt the manuscript was worth publishing. One can see how important the German language was in physics at that time – after all, many of the great physicists came from Germany in those days.

Einstein was impressed by Bose's work and gladly agreed to the request. He translated it and ensured that it was published in the renowned *Zeitschrift für Physik* under the title *Plancks Gesetz und Lichtquantenhypothese* (Planck's Law and the Light Quantum Hypothesis). He added the following remark: "Bose's derivation of Planck's formula appears to me to be an important step forward. The method used here gives also the quantum theory of an ideal gas, as I shall show elsewhere."

Einstein had realized that Bose's method was a universal tool that could be used to describe more than just a gas of photons, which is exactly what electromagnetic heat radiation is in quantum mechanics. Any other ideal gas – whether it be composed of electrons, atoms, or molecules – can also be described quantum-mechanically using this method, because de Broglie's relationship between wavelength and particle momentum also applies to these particles.

When Einstein worked out this idea for a gas consisting of individual atoms, he made a discovery: the equations showed that, at a given temperature, only a certain number of particles with thermal energy can be contained within a given volume. The probabilities that the particles occupy energy states above the energetically lowest quantum state – the ground state – permit only a limited number of particles in these higher energy states. There is a maximum upper limit for the density of particles with thermal energy. The lower the temperature, the lower the upper limit. But what happens if you add still more particles to the given volume?

In his paper *Quantentheorie des einatomigen idealen Gases* (Quantum Theory of a Monoatomic Ideal Gas), Einstein claims that these surplus particles would then all have to accumulate in the ground state. So something similar to the compression of steam should occur if the temperature is kept constant by cooling: if the volume goes below the so-called saturation volume, part of the steam will condense to a liquid. Alternatively, we can leave the volume constant and lower the temperature – this is exactly what happens when water vapor condenses into mist or clouds.

With real steam, the attractive forces between the water molecules cause them to condense into droplets. With an ideal quantum gas there are no such forces between the particles – this is exactly the definition of an ideal gas. Nevertheless, according to Einstein, a condensation of gas atoms in the lowest quantum state should occur at very low temperatures for quantum mechanical reasons. Today this phenomenon is called *Bose–Einstein condensation*.

This quantum condensation should not be understood in the same way as the condensation of water vapor. Bose–Einstein condensation means that more and more particles combine to form a single quantum wave and behave synchronously. They no longer exhibit any random thermal motions and therefore do not carry any heat energy.

One important condition for Bose–Einstein condensation has not yet been mentioned, however: the particles must be *bosons*, i.e., particles with an integer spin, for example spin zero, one, or two. Electrons, protons, and neutrons, on the other hand, are not bosons, but *fermions* – they have half-integer spin, in fact, spin 1/2. For fermions, the *Pauli principle* applies. This was formulated by Wolfgang Pauli at about the same time, and states that each quantum state can contain at most one fermion. This is what leads to the different electron shells in atoms, where the quantum states are gradually filled up from bottom to top with the available electrons.

Bosons are not subject to Pauli's restriction. They have no problem gathering in the quantum mechanical ground state, where they are simultaneously described by a single macroscopic wave function that completely determines their behavior. The bosons agree on all observable properties and move quantum-mechanically "in perfect step".

It took until 1995 for Eric Cornell, Carl Wieman, and Wolfgang Ketterle to succeed in producing such a condensate for the first time, and thereby prove that Einstein was right in the assumption he made. They captured a gas of rubidium atoms in a magnetic trap and cooled it down to 170 nK, i.e., to less than a millionth of a kelvin – a technical masterpiece that was rewarded with the Nobel Prize in Physics in 2001.

Why Does Helium Become Superfluid?

The simultaneous behavior of the bosons in a Bose–Einstein condensate is in many ways reminiscent of the behavior of the superfluid component of liquid helium. Could this component perhaps be a Bose–Einstein condensate, as the German–American physicist Fritz London suspected in 1938? Is the superfluidity of helium a macroscopic quantum phenomenon?

This idea is tempting, especially since almost all helium atoms are actually bosons. Most helium atoms contain two protons and two neutrons in their atomic nucleus, whence they are called helium-4 nuclei. The spins of these particles are oppositely oriented and cancel each other out. The same applies to the two electrons of the atomic shell, so helium-4 atoms have a total spin of zero.

There is only one problem: Einstein's idea applies strictly speaking only to an ideal boson gas in which the individual gas particles behave like tiny, perfectly elastic balls. Apart from hard, elastic collisions, there are no forces between the gas particles in this model. In addition, the gas particles must be so small that their volume does not play any role – the particles may collide in principle, but they may not push each other out of the way.

These ideals are not fully satisfied by liquid helium. The attractive force between the helium atoms is weak, but it is not zero – this is exactly why helium condenses to a liquid at low temperatures, and it has nothing to do with the quantum mechanical condensation to a Bose–Einstein condensate. In addition, the atoms are all close together in liquid helium, so their size will certainly play a role. Can Einstein's considerations still be applied to helium atoms under these conditions? Or do the interactions and size of the helium atoms impede the formation of a Bose–Einstein condensate?

This was one of the questions Richard Feynman addressed in the early 1950s. He found that in many respects the interaction and size of the helium atoms would not prevent them from behaving like particles in an ideal gas. The motion of one helium atom is not particularly hindered by the presence of the other atoms, because these atoms can move to the side and thus clear the way. Such evasive motion essentially only causes more mass to be moved overall than in an ideal gas. So it looks as if one helium atom pushing through the others has more mass than if it could move freely. Therefore, Einstein's considerations, which apply to an ideal Bose gas, do indeed transfer quite well to liquid helium. A Bose–Einstein condensate is thus possible.

When we see this reasoning, it seems quite plausible. Do we really need a genius like Feynman to figure that out? Well, after all, the helium atoms constantly push each other aside, something that does not happen in an ideal gas where there are only occasional brief collisions. Moreover, the weak attractive forces are sufficient to bind the helium atoms into a liquid at low temperatures – liquid helium is not only not an ideal gas, it is not a gas at all! Feynman's idea can therefore be challenged. For this reason, Feynman did not simply put forward his argument "from the gut", but fashioned it out of a detailed quantum mechanical analysis, combined with a lot of physical intuition – typical Feynman. He even used path integrals, the tool he had invented himself, a sign of how immensely useful it is in so many applications of quantum mechanics.

So, using his own quantum mechanical methods, Feynman was able to explain why there is a phase transition between helium I and helium II. In helium II, some of the helium atoms are synchronized in a Bose–Einstein condensate which is described by a single wave function, so that all atoms in this condensate behave in exactly the same way. This condensate comprises only one macroscopic quantum state, i.e., the lowest energy state (ground state). Therefore, the condensate lacks any thermal energy.

In Laszlo Tisza's two-fluid model, this condensate corresponds to the superfluid component. But what about the normal component in this model? Can it also be explained quantum-mechanically?

Quantum Excitations: Phonons and Rotons

Since thermal energy is stored in the normal helium component, it cannot comprise only a single quantum state. Otherwise, all atoms in this component would also behave synchronously, in "quantum equal step", and would not execute any random thermal motion.

The normal component must therefore be a mixture of quantum states with higher energy. In 1941, the Soviet physicist Lev Landau made a first attempt to determine these quantum states. To avoid having to deal with a huge number of helium atoms, he simply described liquid helium as a continuous medium without an atomic substructure, and established a quantum theory for such a substance.

Landau's investigations suggested that there should only be a single type of quantum excitation below about 0.5 K, and that it should be closely related to sound waves in liquid helium. These quantum excitations are called *phonons*. Just as photons are the quanta of light waves, phonons are the quanta of sound waves. In a liquid, the higher the temperature, the more phonons there are. The energy of these phonons increases in inverse proportion to the wavelength of the sound – so long-wave sound consists of low-energy phonons (the same is true for light and photons).

Above about 0.5 K, another type of quantum excitation comes in, which Landau called *rotons*. As the name suggests, rotons are quanta of macroscopic rotational movements in liquid helium. They therefore have nothing to do with the rotation of individual helium atoms, but always involve the motion of many helium atoms. Landau found out that a certain minimum energy is needed to create a roton. This is due to the fact that, in quantum mechanics, angular momentum can only take discrete values – so there is no arbitrarily small angular momentum. However, it was not known how helium atoms could coordinate their motion in a roton. Since Landau had not looked at the individual helium atoms, he could not say anything about that.

On the other hand, Landau's work did provide some explanation as to why the Bose–Einstein condensate – i.e., the group of helium atoms in the common ground state – is superfluid and shows no internal friction: it is just difficult to extract a group of helium atoms from the Bose–Einstein condensate in such a way that a phonon or roton with the appropriate energy and momentum is formed. Friction thus has hardly any point of attack to transfer energy to helium atoms in the condensate, and the superfluid component continues to move without any friction. Only at a certain minimum speed is there enough energy and momentum available to create phonons and rotons – and the superfluidity then collapses.

Phonons and rotons which already exist in the liquid and move freely in it rather like gas atoms can, however, exert friction, just like the atoms in a gas. The normal component of liquid helium consists of just these phonons and rotons. With increasing temperature, more and more phonons exist, and from 0.5 K onward the number of rotons also begins to increase, whereupon the proportion of atoms in the normal component increases, while the proportion in the superfluid component becomes smaller and smaller and finally disappears completely at 2.2 K.

Landau's model, with its phonons and rotons, provided a pretty good explanation for why Tisza's two-fluid model worked so well, and why the superfluid component had no viscosity. But other questions remained unanswered. For example, it was unclear what rotons actually are and what role the quantum-mechanical indistinguishability of helium atoms was playing, a crucial issue for the formation of a Bose–Einstein condensate.

Looking for the Wave Function

The phenomenon of superfluidity is therefore closely related to the fact that, at extremely low temperatures, there are only certain kinds of quantum excitation in liquid helium: phonons and rotons. To lose energy through friction, new phonons and rotons must be created. But this is not so easy, as we have seen.

Feynman asked himself whether there might not be other low-energy quantum excitations. Why should there only be phonons and rotons? In order to make progress on this, Feynman had to take into account the atomic structure of liquid helium in a quantum mechanical way. His aim was to put all the previous phenomenological models on a solid basis and derive them directly from a fundamental quantum mechanical description.

He had set his sights high, because the quantum mechanical properties of a system consisting of many particles, such as liquid helium, can be extremely complicated and could never be described with absolute mathematical precision. So we have to make the right simplifications and we need a good intuitive feeling for what we can and cannot omit. There could hardly have been anyone better equipped than Feynman for this task. As a master of physical intuition, he was used to thinking in pictures. He was never satisfied with abstract mathematical descriptions, but always tried to develop a clear physical idea of what was going on. This would also be an advantage to him here.

It would take us too long to go into the details of Feynman's ideas here, because they remain complex, despite their vividness. But let us at least try to sketch some of his insights.

How could one imagine a quantum description of liquid helium? In classical physics, one would simply specify where each individual helium atom is located and how it moves over time. This is no longer possible quantum-mechanically, because we do not know where the individual atoms are located. We can only give the probabilities for finding certain atoms at certain locations when a measurement is made. These probabilities do not apply to individual atoms, but only to the complete set of all atoms. For example, we can establish how likely it is to find a helium atom at location A and at the same time another helium atom at location B and another at location C, and so on.

Here it does not matter which helium atom is found at which location, because the helium atoms are quantum-mechanically indistinguishable. If two atoms are swapped, the probability must remain the same. This has far-reaching consequences, which are described in more detail in Infobox 4.2.

Infobox 4.2: Pauli exclusion principle and the spin–statistics theorem

If several particles of the same kind come together in quantum mechanics, they are in principle indistinguishable. For example, if we look at the electrons in the shell of an atom, we never know which of them we have just encountered at a certain location. The same applies to helium atoms in liquid helium. If two of these atoms are swapped, the probability of finding the atoms at the corresponding locations remains unchanged.

To determine this probability, the mutual quantum wave of the particles must be squared, as we know from Chap. 1. So there are two possibilities: the quantum wave either remains completely unchanged when the particles are swapped, or it simply changes its sign, in which case wave crests then become troughs and vice versa. The sign disappears when squaring and the probability therefore remains unchanged.

Interestingly, which case we have depends on the spin of the particles. For identical particles with integer spin (bosons, like helium-4 atoms) the sign does not change. For particles with half-integer spin (fermions, like electrons) the sign changes. This relation is called *spin–statistics theorem*.

One extremely important consequence of this relation is that two fermions cannot be put into the same quantum state. Wolfgang Pauli formulated this principle in 1925, and it was called the *Pauli exclusion principle* in his honor.

Since electrons are fermions, two electrons cannot be in exactly the same quantum state in the shell of an atom. In a sense, electrons avoid each other and do not like being in the same place or in the same shell. Thus, as the number of electrons increases, the quantum states in the atomic shell are filled up step by step – in principle, the whole of chemistry is based on this finding.

The situation is completely different for bosons. They like to be together and can even populate the same quantum state in any numbers, as the Bose–Einstein condense so clearly illustrates.

So why does spin influence particle behavior in this way? Why do fermions avoid each other while bosons like to get together?

Richard Feynman makes an interesting comment on this question in Vol. III of his *Feynman Lectures* (Quantum Mechanics). At the end of Chap. 4-1, he writes:

> We apologize for the fact that we cannot give you an elementary explanation. […] It appears to be one of the few places in physics where there is a rule which can be stated very simply, but for which no one has found a simple and easy explanation. The explanation is deep down in relativistic quantum mechanics.

So there is an explanation – Markus Fierz and Wolfgang Pauli worked it out in 1939 and 1940. But the explanation is complex and uses both quantum mechanics and relativity theory.

In 1986, Feynman himself also tried to give an explanation that would be as simple as possible. You can find it in his Dirac Memorial Lecture entitled *Elementary Particles and the Laws of Physics*. But even his explanation is still far too complex to be presented here.

In order to describe liquid helium quantum-mechanically, we must in principle solve the corresponding Schrödinger equation. This is already difficult even for just a few particles, and with trillions of trillions of particles in a liquid, it becomes completely impossible.

Feynman therefore began to build up an idea of what the quantum description might look like on the basis of general physical considerations. How could one imagine the excited quantum states (in short, quantum excitations) with low energy – the phonons and rotons? How did helium atoms typically arrange themselves in liquid helium when they had little energy available?

The helium atoms will have the lowest energy when they move very slowly, and when they remain at an equal mean distance from each other, so that the attractive and repulsive forces between them are balanced. These configurations correspond to the quantum mechanical ground state, i.e., the superfluid component.

What happens when a little bit more energy is available? Feynman argued that the atoms then move closer together in some places and further apart in others. The density of the liquid helium therefore varies slightly from place to place. Such configurations correspond to sound waves that propagate through the liquid. The longer the wavelength of these sound waves, the lower their energy. So areas with different densities must be far apart, if only little energy is available.

Based on such considerations, Feynman developed a quantum description of liquid helium. With this method, he was able to visualize how phonons are formed as sound wave quanta at low energies. Thus Feynman confirmed Landau's assumption that phonons are the quantum excitations with the lowest energy, putting it on a solid foundation.

We cannot go into the details here, but one aspect proved to be extremely important in Feynman's considerations: the indistinguishability of helium atoms. He had to take into account the fact that the probability must not change if two helium atoms were swapped – otherwise his considerations would not work. As bosons, helium-4 atoms like to be in the same quantum state. This was something Feynman needed to bear in mind in order to understand the quantum behavior of liquid helium.

Landau had called for a second type of quantum excitation, which required a certain minimum energy to be produced: the roton. These had to correspond to other configurations of the atoms where the mean density did not change on a large scale, because this case was already covered by the phonons. Feynman explored the different possible configurations with the help of many vivid representations, in which the indistinguishability of the helium atoms played a central role, and gradually built up useful ways to describe the different quantum excitations.

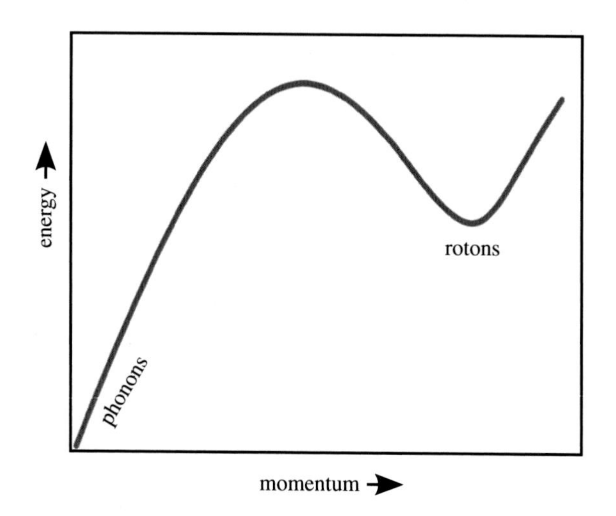

Fig. 4.9 Schematic relationship between the energy and momentum of quantum excitations in superfluid helium

He even succeeded in establishing a formula called a dispersion relation for the relationship between the energy and momentum of the quantum excitations, thus making it clear how phonons and rotons differ from each other. The only freedom in this formula was a certain function related to the mean distances between the helium atoms in the liquid. This function could be determined experimentally by scattering X-rays or neutrons on liquid helium. At low energies and momenta, the energy of the quantum excitations increases linearly with the momentum, just as in the case of a phonon. For higher values of the momentum, on the other hand, the typical mean distance between the helium atoms in the liquid produces a trough with a minimum energy, and that is the energy–momentum range where the rotons are located (Fig. 4.9).

At first, Feynman could only vaguely guess how to visualize rotons in concrete terms. The situation with phonons was clearer: they corresponded to weak large-scale density fluctuations in the liquid helium, involving a large number of atoms.

The excitations with higher energies to the right of the roton trough in Fig. 4.9, on the other hand, had rather to be assigned to the oscillations of individual atoms in the force field of their neighboring atoms. Rotons were somewhere in-between, i.e., some atoms move together, swinging or rotating collectively.

All these quantum excitations taken together formed what was called the normal component in the two-fluid model. That closed the circle and Feynman thus got back to the quantum-mechanical model that Landau had already found for the two-fluid model.

In his work on rotons, Feynman came across an interesting possibility: tiny ring-shaped vortices could form in liquid helium, similar to those known from smoke rings. The helium atoms flow through the inside of the vortex ring and then back around the outside. Feynman and his doctoral student Michael Cohen suspected that rotons could be a kind of quantum version of these micro vortex rings, and constructed an improved roton wave function with this idea. They succeeded in reproducing the experimentally determined value of the minimum roton energy quite well. Feynman's older results had been too high there.

Michael Cohen was one of the few Ph.D. students Feynman had who was able to work successfully with his supervisor. Other doctoral students, in contrast, often had difficulties with him, as he could hardly bear to leave an interesting problem to someone else – he was simply too tempted to solve the problem himself. Cohen had therefore approached things differently: he had studied Feynman's work on liquid helium in great detail and then came to Feynman with his own ideas. In this way a fruitful collaboration developed between Feynman and Cohen.

Despite all his attempts, Feynman failed to explain one phenomenon of low-temperature physics: superconductivity, i.e., the complete disappearance of electrical resistance in very cold electrical conductors. Although superconductivity is similar to superfluidity in many ways, he could not find a satisfactory explanation for this. The American physicists John Bardeen, Leon Neil Cooper, and John Robert Schrieffer succeeded in doing this in 1957 – their so-called BCS theory was rewarded with the Nobel Prize in Physics in 1972. The basic idea can be described as follows.

As fermions, the electrons of the electric current are actually unable to form a Bose–Einstein condensate that flows through the electrical conductor without electrical resistance (the electrical counterpart to mechanical viscosity). Therefore, the electrons must first combine to form what were called Cooper pairs, in which the quantized oscillations of the atomic lattice played an important role as mediators. In this way, the spins of the electrons neutralized each other in the Cooper pairs and they could populate the common ground state as bosons. Cooper pairs could subsequently move in a completely uniform manner, like the helium-4 atoms in the superfluid liquid component. In both cases, energy loss due to excitation of other quantum states is suppressed, so that the current of Cooper pairs flows just as smoothly as the stream of helium atoms.

4.3 Right and Left: The Violation of Mirror Symmetry

During the years in which Feynman was working intensely on the properties of liquid helium, he continued to keep an eye on current developments in particle physics. However, since he no longer actively participated in this field of research himself, he soon had the feeling that he was not really quite up to scratch: "I was always a little behind. Everybody seemed to be smart, and I didn't feel I was keeping up."[5]

But the events of 1956 would show that he was mistaken. Feynman was already 38 years old at the time and was about to make a discovery that he personally considered to be one of his greatest successes. It all began with the *International Conference on High Energy Physics (ICHEP)*, which was held for the sixth time in Rochester in the northwest of New York State in April 1956. At this so-called *Rochester Conference*, the latest results in particle physics have been presented and discussed from 1950 until today, although the conference location has been changing constantly since 1958. In 1956, the focus was on a phenomenon that was causing great confusion: the so-called τ-θ puzzle (pronounced "tau theta puzzle").

This puzzle concerns the decay of a certain group of mesons that were discovered in cosmic rays in 1947. As we know today, mesons contain a quark and an antiquark, while protons and neutrons consist of three quarks. The mesons discussed in Rochester at that time are now called K mesons or simply *kaons*. They have about half the mass of a proton or neutron and, rather like the lighter pions, come in three variants: positively charged K^+, negatively charged K^-, and electrically neutral K^0 (Fig. 4.10).

The first surprise encountered with the kaons was their unusually long lifespan. The K^+ decays on average after about 12 ns (1.2×10^{-8} s). This may seem very short to us, but a beam of light can travel 3.6 m in this time. The somewhat heavier ρ mesons decay after just 4.5×10^{-24} s – two and a half trillion times faster; a beam of light cannot even pass through a single proton in this extremely short period of time.

[5]From *Surely You're Joking, Mr. Feynman!*

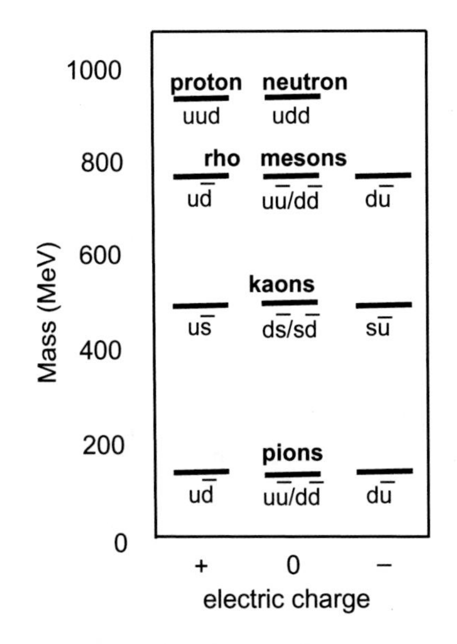

Fig. 4.10 Masses, charges, and quark contents of some mesons. The proton and neutron are also shown for comparison. Antiquarks are denoted by a bar over the symbol for the corresponding quark

Why does the K^+ live so much longer? In order to understand this, we first have to consider why particles decay at all. Why does a pion, kaon, or ρ meson decay while a proton is stable?

Why Most Particles Decay – And Some Do not

At first glance, there seems to be no reason why a proton should not decay. The energy stored in its mass could, for example, be released to a large extent by the decay into a light positron and some photons. Charge, energy, and momentum could be maintained. However, stable matter as we know it would then be impossible. So there must be some criterion that decides whether a particle is stable or decays into lighter particles. From what was said in Chap. 3 we can formulate this criterion quite simply in the following way: *there must be a Feynman diagram for the decay!*

Feynman diagrams generally contain two building blocks: lines that represent the motion of individual particles in space and time, and vertices where new particles are created or destroyed. Each fundamental interaction has its own vertices, characteristic of that particular interaction.

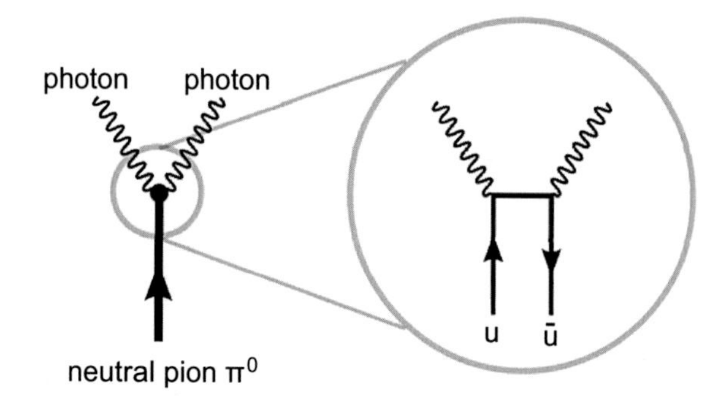

Fig. 4.11 Electromagnetic decay of the neutral pion π^0 into two photons. Inside the pion, an up quark and an up antiquark destroy each other

We have already got to know one example in detail: the electromagnetic interaction. At its vertices, a charged particle emits or absorbs a photon. The charged particle itself does not change its identity – for example, an electron remains an electron. Its line only gains a kink at vertices in such a diagram. However, the particle can change its time direction, i.e., it can be thrown back into the negative time direction in the diagram and thus become an antiparticle. In this way, Feynman diagrams can be constructed in which a particle and its antiparticle destroy each other and radiate photons (see Fig. 4.11).

So far everything is clear for the electromagnetic interaction. But what happens with the other three interactions we know about: gravity and the strong and weak interactions? Fortunately, we can keep gravity out of the picture. Gravity is so weak that it plays no significant role as a force between individual elementary particles.

We must, however, take into account the strong interaction. It is even stronger than the electromagnetic interaction and holds the protons and neutrons together in the atomic nucleus. But what do the Feynman diagrams of this interaction look like? Is there something like a photon that can be emitted and absorbed at the vertices? Today we know that this is actually the case: in the strong interaction, the photon is replaced by the *gluon*, which couples to a quark at a vertex.

With the help of the corresponding Feynman diagrams, for example, the decay of the ρ meson into two pions can be well explained (Fig. 4.12). And because, as its name suggests, the strong interaction is the strongest of all interactions, the decays it mediates are also especially fast – about ten million times faster than the electromagnetic decays. This explains the very

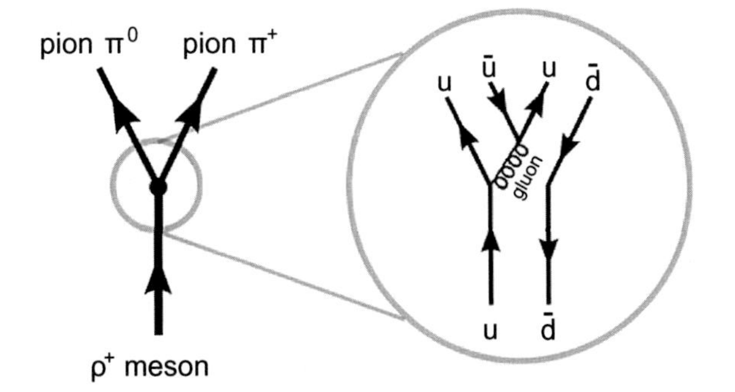

Fig. 4.12 In the very fast decay of the ρ^+ meson into two pions due to the strong interaction, a quark emits a gluon, which immediately transforms into a quark and an antiquark. The two quarks and two antiquarks are then grouped into two pions. Note that gluon lines are "curled" and not wavy

short lifetime of the ρ meson, which decays almost at the very moment of its creation.

Strangeness Makes Particles Live Longer

Now we are ready to reconsider the kaons and the question of why the K^+ lives so much longer than the ρ meson, even though both mesons often decay into pions. With our current knowledge of the quarks inside the mesons, the answer is simple: the K^+ contains a so-called strange antiquark, which has to transform into an up antiquark upon decay. But neither the strong nor the electromagnetic interaction are able to do this, because they cannot change the type of a quark. So the K^+ cannot decay through these two interactions.

In the early 1950s, quarks and gluons were still unknown. The American physicist Murray Gell-Mann and his Japanese colleague Kazuhiko Nishijima therefore had to solve the problem in a different way. They assumed that there must be a new quantum number, to which Gell-Mann gave the suggestive name *strangeness* – particles with strangeness are also called *strange particles*. Nishijima had used the term *eta-charge* instead – not a very lucky choice, because this somewhat uninspiring term had no chance compared to "strangeness". Gell-Mann was famous for introducing this kind of exotic, but at the same time catchy term, which quickly stuck in physicists' minds. He also later coined the word *quark*, introduced in 1964 (see Sect. 6.1).

Gell-Mann now stipulated that neither the strong nor the electromagnetic interaction should be able to change the total strangeness of a particle. It then suffices to assign a strangeness +1 to the K^+ and zero to the pions, and the K^+ can no longer decay into pions via these interactions. Of course, the mysterious strangeness is nothing other than the number of strange antiquarks minus the number of strange quarks, but Gell-Mann and Nishijima could not yet have known that.

But then why can the K^+ decay at all?

If only the electromagnetic and the strong interaction existed, the K^+ would actually be a stable particle. In order for the K^+ to decay, another interaction had to be involved. This is called the *weak interaction*, because the decays it causes occur much more slowly than those due to the strong or even the electromagnetic interaction (strength and speed go hand in hand in this case).

The weak interaction does not care about strangeness, because it can convert one type of quark into another. In this way, it enables a multitude of decays that would otherwise be impossible. For example, it can convert an up quark into a down quark and vice versa, and this is the basis of the beta decay – more about this later. A strange antiquark can also turn into an up antiquark, and this is exactly what happens when a K^+ decays into pions (Fig. 4.13).

The weak interaction can do even more: it can transform an electron into an electron neutrino and vice versa, or it can create an electron together with its antineutrino. Analogously, it can create the heavy counterpart of the electron, the muon, together with the associated muon neutrino, or its superheavy counterpart, the tau, together with the associated tau neutrino. However, it

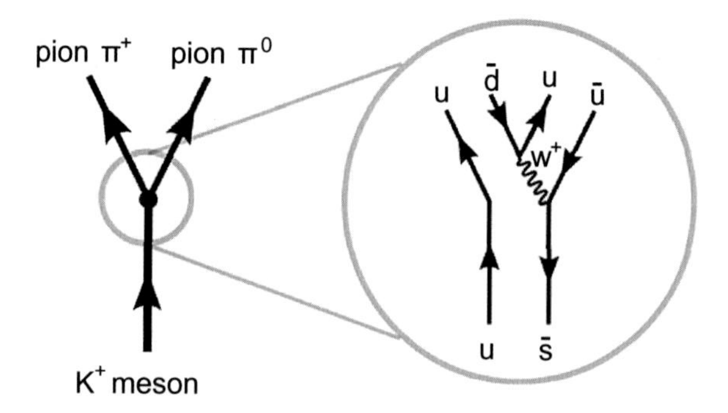

Fig. 4.13 Decay of the K^+ meson into two pions by the weak interaction. In the quark picture on the right, the quark type changes at the W^+ vertices

must always be the associated neutrino that gets created, simply because neutrino types are in fact distinguished from one another via these processes. By the way, the weak interaction will never turn a quark into a lepton, i.e., an electron, muon, tau, or neutrino – it must produce another quark.

It would take until the end of the 1970s before physicists obtained a detailed understanding of all the various relationships between quarks, leptons, and the strong, weak and electromagnetic interaction. The corresponding theory carries the unimpressive name *Standard Model of particle physics* or simply *Standard Model.* But this rather drab name is deceptive, because the Standard Model is really a jewel – one of the best theories that modern physics has ever created. A short summary can be found in Infobox 4.3, and in Chap. 6 we shall return to the Standard Model – although we shall no longer focus on the weak interaction there, but rather on the quarks and their strong interactions.

Infobox 4.3: The Standard Model of particle physics

The Standard Model was developed in the 1970s and describes all known particles and their interactions with the exception of gravity. All predictions of the Standard Model have since been confirmed to a high level of accuracy.

In the Standard Model, the basic building blocks of matter are six quarks and six leptons (electrons, neutrinos, and their "big brothers"). Three types of interactions between these particles are possible: strong, weak, and electromagnetic interactions (Fig. 4.14). These interactions are mediated by elementary bosons – the photon, the gluon, two charged W-bosons, and a neutral

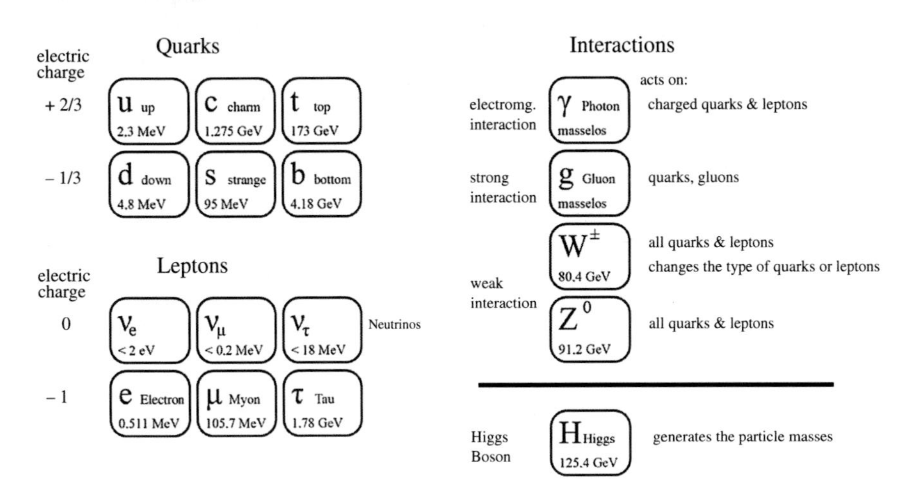

Fig. 4.14 The particles in the Standard Model of particle physics and their masses (in mega- and gigaelectronvolts)

Z-boson – which couple to the quarks and leptons at vertices in the Feynman diagrams.

The massless gluons only interact with quarks or other gluons. They bind quarks together to form baryons (e.g., protons and neutrons) or quarks and antiquarks to form mesons – there are no free quarks or gluons. The photons, which are also massless, couple to all electrically charged particles, and the very heavy W bosons and Z boson couple without exception to all quarks and leptons. The W bosons convert different types of quarks or leptons into one another and thus mediate a great many weak decays. But note that a quark is never converted into a lepton or vice versa.

The final ingredient required for the Standard Model is the so-called Higgs particle, which is responsible for the masses of the W and Z bosons as well as those of the quarks and leptons. As early as 1964, Peter Higgs, François Englert, Robert Brout, and others had predicted the existence of what became known as the Higgs particle. It was a great triumph for the Standard Model when it was actually found in 2012.

So Gell-Mann and Nishijima could successfully explain the slow decay of the K^+ with their strangeness rule: it could only decay through the weak interaction. This was already well known at the Rochester Conference in 1956. The so-called τ-θ puzzle concerned a peculiarity of the K^+ decay, which became the subject of intense discussion at this conference.

The τ-θ Puzzle: Two Particles or just One?

The K^+ can decay in several different ways by the weak interaction. For example, two pions can be created, but sometimes also three pions:

$$K^+ \to \pi^+ + \pi^0$$
$$K^+ \to \pi^+ + \pi^0 + \pi^- \quad \text{or} \; \pi^+ + \pi^0 + \pi^0$$

We may note immediately that electric charge conservation is guaranteed in each case. The problem with these decays is much more subtle: it only arises if we assume that the laws of nature do not depend on whether we look at an experiment in a mirror or not. If we take mirror symmetry for granted in nature, then a physical process viewed in a mirror should look exactly like a real process. According to the commonly held view at the time, the laws of nature should indeed be mirror symmetric, and we should not be able to decide in principle whether an experiment is being viewed in a mirror or not.

This assumption seems perfectly plausible. Why should nature distinguish between right and left? One seems just as good as the other. In addition, the electromagnetic interaction had already been proven to satisfy this assumption. Why should things be different with the weak interaction causing the above decays of the K^+?

If one assumes that the laws of nature are mirror symmetric, the following result can be derived from quantum mechanics: a particle that decays into two pions cannot decay into three pions and vice versa. Unfortunately, some effort is required to justify this statement, and we shall not be able to go into the details here. We just have to bear in mind that, according to the law of mirror symmetry, it is simply forbidden that something able to go into two pions can also go into three pions. This is also known as the *parity rule*.

In order to resolve the contradiction here, it was assumed at that time that there had to be two different particles, instead of just the K^+. One of them, called θ^+ (theta-plus), would disintegrate into two pions, while the other, called τ^+ (tau-plus), would decay into three pions. The only problem was that otherwise the two particles had absolutely identical properties: the same mass, lifetime, and so on. Everything suggested that they were one and the same particle, the one we call K^+ nowadays. But then the laws of nature cannot be mirror symmetric. This was once again unthinkable for the great majority of physicists – and especially for the theorists. Something as beautiful and obvious as mirror symmetry seemed untouchable.

So that was the general situation in which the particle physics community came together at the Rochester Conference in 1956. Feynman was also there – he shared a room with the experimental physicist Martin Block. In *Surely You're Joking*, he describes how Block asked him one evening: "Why are you guys so insistent on this parity rule? Maybe the τ and θ are the same particle. What would be the consequences if the parity rule were wrong?"

Are the Laws of Nature Mirror Symmetric?

Feynman replied that there would then be a way to use the laws of nature to define an absolute right and left – they would no longer be on an equal footing. He did not really know whether this would be such a major disaster, so he encouraged his colleague to ask the experts at the conference the next day. But Block feared that no one would even listen to him if he asked such a heretical question, so he asked Feynman to take over. The next day Feynman therefore began to ask around: "I'm asking this question for Martin Block: What would be the consequences if the parity rule was wrong?"

Murray Gell-Mann often teased him about it later on, saying Feynman didn't have the nerve to ask the question for himself. But Feynman simply didn't want to take credit for an idea that someone else had raised, and that might have been of central importance.

The two young Chinese–American physicists Tsung-Dao Lee and Chen Ning Yang, then experts in this field, responded to Feynman's question with a rather complicated explanation that Feynman felt he did not understand particularly well. After all, he had the impression that the issue had by no means already been decided. And like almost everyone else, he considered it unlikely that the parity rule would actually turn out to be violated, and that the beautiful mirror symmetry in nature would not therefore apply.

When Norman Ramsey, an experimental physicist, asked him if he should do an experiment to look for parity law violation, Feynman replied: "The best way to explain it is, I'll bet you only fifty to one you don't find anything." With this opinion Feynman was in good company. Wolfgang Pauli also stated that he could not believe God was a weak left-hander, where the word *weak* referred of course to the weak interaction.

Maybe that discouraged Ramsey, because he never did the experiment – and that turned out to be a mistake! But the theorists Lee and Yang took the question seriously. Lee was convinced that progress could only be made in physics if theory and experiment worked closely together. His credo was: "Without experimentalists, theorists tend to drift. Without theorists, experimentalists tend to falter."[6]

Might the experiments on mirror symmetry yield some kind of surprise? Lee and Yang were wondering if this could be true, and started searching all the available experimental data for deviations from the parity rule. Regarding the strong interaction, they found clear indications that the rule was valid, but for the weak interaction, such indications were missing. Apparently, there was no experimental basis for assuming that the rule of parity would also apply to the weak interaction. Everyone believed in it, but the experiments gave no support for this belief.

[6]Quoted in Alexander Lesov: *The Weak Force: From Fermi to Feynman*, http://arxiv.org/pdf/0911.0058.pdf.

Fig. 4.15 Chien-Shiung Wu (1912–1997) next to Wolfgang Pauli (1900–1958) (© CERN; http://cds.cern.ch/record/42739)

Lee and Yang were determined to answer the question one way or another by carrying out new experiments. So they began to make suggestions and talked to different experimental physicists about how this could be done. Among them was the Chinese–American physicist Chien-Shiung Wu (Fig. 4.15). Madame Wu, as she was also called in reference to the famous Madame Curie, had considerable experience in the field of radioactive beta decays of atomic nuclei. One of Lee and Yang's suggestions was based on just such a beta decay, so Chien-Shiung Wu gladly took up their idea.

We already got to know something about beta decay in Chap. 2. Like the K^+ and π^+ decays, it is mediated by the weak interaction. In beta decay, a neutron in a radioactive atomic nucleus is converted into a proton, emitting a high-energy electron and an electron antineutrino. Today we know what happens in detail in beta decay. Inside the neutron, the weak interaction transforms a down quark into an up quark, thereby turning the neutron into a proton. The energy released in this process is used by the weak interaction to create the electron and its antineutrino according to Einstein's formula $E = m \cdot c^2$ (Fig. 4.16).

The high-energy electrons emitted by an atomic nucleus during beta decay are also known as radioactive beta radiation. They can easily be detected using simple measuring instruments. In contrast, the ghostly antineutrinos, which are also released, almost always remain invisible.

The experiment planned by Madame Wu was based on the radioactive beta decay of cobalt-60 atomic nuclei, which are converted into nickel-60

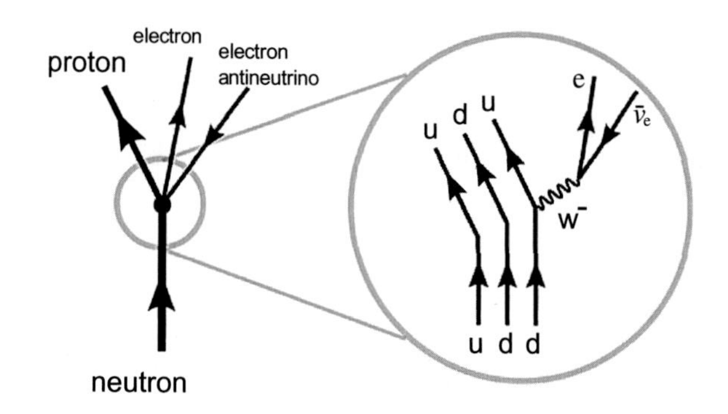

Fig. 4.16 Beta decay of the neutron occurs via the weak interaction. As always, the arrow of the antineutrino is oriented against the time direction, i.e., downwards

nuclei with a half-life of about five years. The question Wu wanted to clarify was this: does this decay respect the parity rule in the sense of being mirror symmetric, i.e., does a mirror-inverted measurement setup lead to a mirror-inverted measurement result?

To answer this question, it is crucial to know that cobalt-60 nuclei have spin, i.e., quantum mechanical angular momentum. In order to know how the nuclei are spinning in the experiment, their axes of rotation must first be aligned parallel, so that all nuclei rotate around the same spatial axis with the same direction of rotation. This can be achieved by using a strong magnetic field, because the nuclei behave like tiny compass needles due to their rotation, aligning themselves parallel to the magnetic field lines.

There is only one problem. The random thermal motion of the atoms constantly disturbs this orientation. Extremely low temperatures of no more than 0.01 K are therefore required to prevent this. Only a few laboratories in the world could reach such low temperatures, among them the cryogenic laboratory at the National Bureau of Standards in Washington, D.C. So Wu went there to carry out her experiment with the help of the resident cryogenic experts.

The key features of the experimental setup can be described as follows. To generate the magnetic field, we require a current-carrying coil, which we imagine oriented vertically so that the magnetic field inside it is also oriented vertically (Fig. 4.17). The spin axes of the cooled cobalt-60 nuclei inside the coil will then also be aligned vertically. The direction of the current in the coil and thus the orientation of the magnetic field should be selected in such a way that the atomic nuclei rotate counter-clockwise when viewed

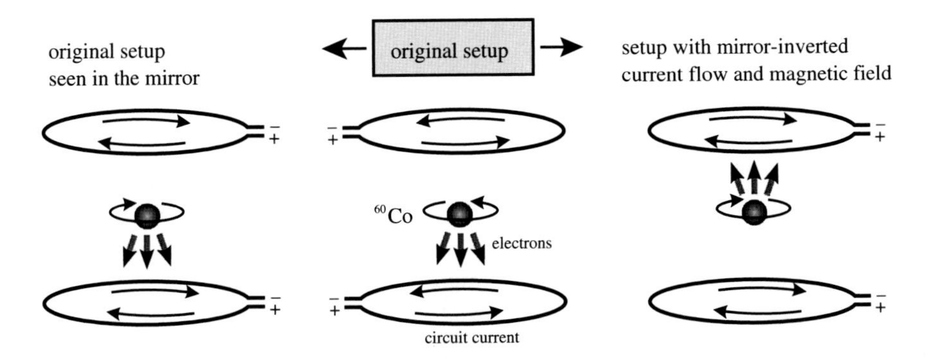

Fig. 4.17 Schematic view of the Wu experiment. The current-carrying coil is indicated by a circuit current at the top and bottom. The original setup is shown in the center. The result of the mirror-inverted experiment is shown on the right, while on the left we see the original experiment when it is actually viewed in a mirror

from above. The aim now is to determine how many electrons are emitted upwards and downwards during beta decay of the nuclei.

The Weak Interaction Violates Mirror Symmetry

What was found was a major surprise: the majority of the electrons were emitted downwards by the rotating cobalt-60 nuclei (Fig. 4.17 center). If we imagine that the curved fingers of our left hand indicate the direction of rotation of a cobalt nucleus, then the electrons would in preference be emitted in the direction of the outstretched thumb pointing downwards.

What happens if we set up the mirror-inverted experiment, as shown on the right of Fig. 4.17? To do this, we could use a mirror-inverted coil to generate the magnetic field. In reality, of course, one simply takes the original coil and reverses the current direction to invert the direction of the magnetic field. This also causes the rotation of the cobalt nuclei to reverse, so they now spin clockwise when viewed from above. Here we see the mirror symmetry of the electromagnetic interaction, which is responsible for the direction of rotation of the nuclei in the magnetic field: a mirror-inverted magnet also produces a mirror-inverted alignment of the rotation of the cobalt nuclei.

Now where are the electrons emitted in preference during beta decay in this setup? Of course, in the other direction, i.e., upwards, as shown by our left-hand rule (Fig. 4.17 right). It makes no difference how the nuclei spin – the preferred direction of the electrons must always be oriented in the same way as the spin of the nuclei which emit them. Thus, the mirror-inverted setup does not give the same result as would be seen by looking at the original setup in a mirror (Fig. 4.17 left). In the mirror-inverted setup the electrons go up in preference, while in the original setup the electrons tend to go down, regardless of whether viewed in the mirror or not, because their direction of motion is the same in the mirror image. We can use this to decide whether we are looking at the original experiment in a mirror or whether we are looking at a mirror-inverted experiment.

Matter and Antimatter: That Small Difference

At this point we can make an interesting thought experiment. Suppose that on a distant planet everything consisted of antimatter instead of matter. Instead of electrons and antineutrinos, we would be dealing with positrons and neutrinos, and the cobalt-60 anti-nuclei would consist of antiprotons and antineutrons. For the inhabitants of such a planet, who themselves would consist of antimatter, all this would be quite normal – for them antimatter would be nothing special. As long as antimatter does not meet matter, it is just as stable as normal matter.

If on this planet Madame Anti-Wu had the idea of carrying out the same experiment as Madame Wu – only with antimatter – what would she find? The electromagnetic interaction gives exactly the same physical results for antimatter as for matter, so the cobalt-60 anti-nuclei would have exactly the same direction of rotation as the cobalt-60 nuclei. However, this time they would emit positrons instead of electrons – but in which direction?

Since this experiment cannot actually be carried out, for obvious reasons, we must derive the result with the help of other experiments and additional theoretical considerations. It turns out that positrons would in preference be emitted upwards, i.e., in exactly the opposite direction to the electrons emitted by the cobalt-60 nuclei. A right-hand rule therefore applies to positrons in the beta decay of cobalt-60 anti-nuclei. Thus, antimatter behaves in a mirror-inverted way compared to normal matter in weak decays.

If there were no weak interaction, it would be impossible to distinguish between matter and antimatter. If all matter in the universe were replaced by antimatter, this would make no difference to the laws of nature, for the concepts of positive and negative electric charge are interchangeable and completely arbitrary. This is called C-invariance or C-symmetry, where C stands for the term *charge conjugation*.

Weak interaction violates this C-invariance, because in beta decay, for example, the antimatter universe would behave in a mirror-inverted way compared to the matter universe. But it was generally thought that, if we looked at the antimatter universe in a mirror, we wouldn't see any difference in the laws of nature. This law is called *CP invariance*, where P stands for *parity*.

As James Cronin, Val Fitch, and others discovered eight years later (in 1964), the law of CP invariance is not absolutely valid either. There are tiny deviations from this rule in the weak decays of neutral K^0 mesons and some other particles. It is not so easy to see why, so we will not go into the details here. What is essential is that CP invariance is also violated in nature: an antimatter universe viewed in the mirror is not absolutely identical in all respects to a matter universe. However, we would have to look very closely and measure very accurately to find any difference.

The fact that matter and antimatter differ slightly is very important for our universe. In 1967, the Russian physicist and later Nobel Peace Prize winner Andrei Sakharov found that, without this small violation of CP invariance, exactly the same amount of matter and antimatter would have been produced in the Big Bang. In Chap. 3 we saw that a small asymmetry of one billionth is needed to explain how matter can be present in the universe today. Fortunately, this small difference between matter and antimatter does indeed exist, otherwise we would not be here today.

So can we find an invariance which we believe today to be fully valid in nature? The answer is affirmative. However, not only do we have to replace matter with antimatter and take a mirror image of the world, but we must also let time run backwards. Then no difference should be visible for the fundamental laws of nature. This basic insight is called *CPT invariance*, where T stands for *time*. Time reversal comes into play because, according to Feynman, a particle that moves backwards in time corresponds to an antiparticle that moves forward in time. CPT invariance had only been demonstrated theoretically about a year before the conference (i.e., in 1955) by Wolfgang Pauli, and independently by Gerhart Lüders, from very general assumptions in quantum field theory. At that time Pauli was already 55 years old, so it was not such a bad performance for a physicist of this age!

At the time Pauli was working intensively on these symmetries, and was one of those most surprised by the discovery of parity violation – i.e., the violation of mirror symmetry. Why should the weak interaction dance so far out of step when the parity rule had proved to be absolutely valid for the strong and electromagnetic interactions? What did mirror symmetry have to do with the strength of an interaction?

Even today, no one knows the answer to these questions. The violation of mirror symmetry in weak interactions seems like a joke of nature. Wolfgang Pauli felt the same way and wrote the following obituary when the results of the Wu experiment became known:

> It is our sad duty to make it known that our dear female friend of many years PARITY gently passed away on January 19, 1957, following a moment of brief suffering caused by its experimental treatment. On behalf of the bereaved e, μ, ν.

The bereaved were of course the electron, muon, and neutrino, which play an important role in many weak decays. As we see again, hardly any of the great physicists of the day had seriously considered the possible violation of mirror symmetry. It had always seemed so natural, so self-evident. Clearly, our intuition can sometimes be a poor advisor when it comes to the laws of nature.

Lee and Yang: Neutrinos Are Left

Soon after the results produced by Madame Wu, parity violation was also identified in other weak decay processes. Lee and Yang enthusiastically welcomed these results and published their first theoretical explanation in March 1957, under the title *Parity Nonconservation and a Two-Component Theory of the Neutrino*. In December of the same year, they received the Nobel Prize in Physics for their groundbreaking study of the rules of parity. Madame Wu, however, came away empty-handed, although she certainly deserved to be considered as well.

What Lee and Yang suggested with their *two-component theory of the neutrino* was that the main cause for parity violation in weak decays was in fact to be found in the neutrinos. Neutrinos can only be influenced, created, and destroyed by the weak interaction, and they are the only such particles, so they should reflect the peculiarities of this interaction particularly clearly. To explain parity violation, Lee and Yang put forward the following hypothesis:

> Neutrinos always rotate in the same direction around their flight axis.

Now it is not at all easy to prove the validity of this statement. Neutrinos don't carry an electrical charge, and neither are they subject to the strong

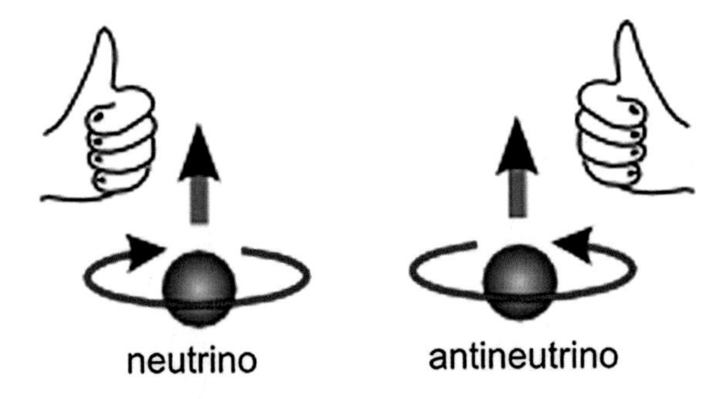

Fig. 4.18 Neutrinos always spin in a left-handed sense and antineutrinos in the right-handed sense around their direction of motion

interaction. That's why we don't notice, for example, that our planet, and indeed we ourselves, are continually being bombarded by neutrinos: over 60 billion of them pass through every square centimeter each second. Most of these neutrinos originate from nuclear fusion processes inside of the Sun. They interact so little with matter that they can pass through the Earth almost unhindered. Only very rarely will one of them interact with an atomic nucleus and thus become noticeable.

Since neutrinos have such ghostly behaviour, there was at first some doubt as to which way they spin. It was only one year earlier, in 1956, that these elusive particles had been detected in experiments for the first time. But it soon became clear that they all spin to the left when viewed in the direction of flight. Maurice Goldhaber and his colleagues proved this experimentally in the same year (Goldhaber experiment). So if the thumb of our left hand points in the flight direction, a neutrino always spins in the direction of our curved fingers. Put another way, all neutrinos are left-handed – or again, right-handed neutrinos do not occur in nature (Fig. 4.18). And of course, it's the other way around for antineutrinos.

Strictly speaking, this statement only makes sense if neutrinos are massless and move at the speed of light like photons. If they had a notable mass, they would have to travel more slowly than light and could be overtaken by a fast observer. For this observer, the neutrino would fly in the opposite direction and he/she would see a neutrino rotating the other way around its flight axis. One observer would therefore see a left-handed neutrino and the other a right-handed neutrino, so the statement that all neutrinos are left-handed would no longer make sense. It is precisely for this reason that there must be left-handed electrons as well as right-handed electrons, because electrons have a mass and can be overtaken.

Mathematically, the masslessness of neutrinos makes it possible to describe them in the Dirac equation by a wave with only two components, whereas four components are needed for the electron. For a particle that is always left-handed, you only need two components, whereas a massive particle like the electron can also be right-handed, so you need two more components. This is exactly what Lee and Yang meant when they spoke of a two-component theory for the neutrino.

As far as we know today, neutrinos are not really completely massless. Their masses are difficult to measure and must be extremely small, but they definitely have a nonzero value. Nonetheless, in almost all processes, neutrinos can be treated as massless particles in very good approximation. So the chances of finding a right-handed neutrino are exceedingly small and we will not take the tiny neutrino masses into account in the following, assuming that they are in fact completely massless.

With Lee and Yang's idea, we can explain why the majority of electrons in the Wu experiment are emitted downwards. In radioactive beta decay, a cobalt-60 atomic nucleus turns into a nickel-60 atomic nucleus, which spins less than the parent cobalt-60 nucleus. The cobalt-60 nucleus has spin 5, while nickel-60 only has spin 4. Therefore, one unit of quantum mechanical angular momentum must be transferred to the electron and its antineutrino during decay. Since these particles are fermions, in fact, spin 1/2 particles, they can only carry half a unit of angular momentum each, whence the angular momentum unit is distributed evenly between them. The electron and antineutrino thus rotate counter-clockwise when viewed from above, with half an angular momentum unit each.

During beta decay, the electron and its antineutrino are emitted in approximately opposite directions for reasons of momentum conservation. The particle flying upwards must therefore rotate right-handedly around its flight direction, while the particle flying downwards must rotate left-handedly. Since an antineutrino can only rotate in a right-handed manner, it must move upwards, and the electron thus has no other choice than to move downwards (Fig. 4.19).

Since neutrinos are involved in many weak decays, this scheme can often be successfully applied to explain parity violation. However, if there are no neutrinos in the game, as when the K^+ decays into two or three pions, this explanation does not work. So the parity violation in the τ-θ puzzle cannot be explained in this way, as Lee and Yang themselves noted in their paper. Such an explanation would only be possible within a fundamental theory of weak interactions that applied to all decays. The only available theoretical description, proposed by Enrico Fermi in 1933 and which had always been

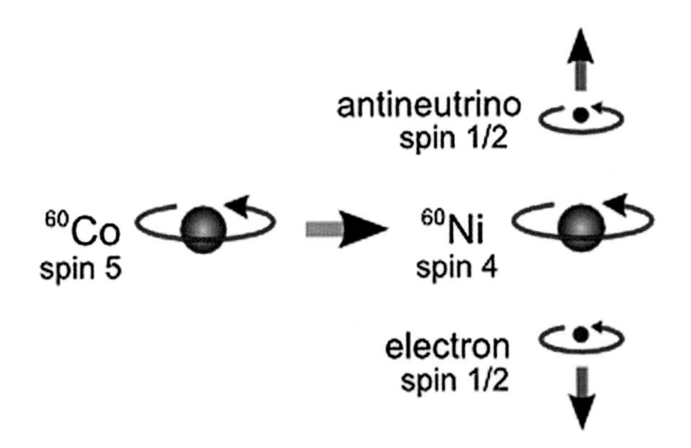

Fig. 4.19 Spins of the particles in the Wu experiment

used until then, could not achieve this, since it did not yet take parity violation into account. Unfortunately, Fermi, whom Feynman knew from Los Alamos, could no longer take part in the theoretical debate himself – he had passed away in November 1954, i.e., around two years earlier, at the age of only 53.

Feynman: The Weak Interaction Prefers Left

At the next Rochester Conference in April 1957, Lee and Yang presented the ideas from their recent publication. Feynman also attended the conference. This time he stayed with his sister Joan in nearby Syracuse and took the paper with him to her home: "I can't understand these things that Lee and Yang are saying. It's all so complicated," he complained to her. Then Joan gave him the same advice *he* had given her a long time ago when he had given her an astronomy textbook for her fourteenth birthday: "What you should do is imagine you're a student again, and take this paper upstairs, read every line of it, and check the equations. Then you'll understand it very easily."

He took her advice, and lo and behold – it was actually quite simple. Feynman had only let himself be intimidated because he believed he had long since been left behind by the others in particle physics. The description of the neutrino by a wave function with two instead of four components reminded him of his own attempts to derive the Dirac equation in his path integral formalism. He had encountered difficulties and, like Lee and Yang, had begun to experiment with two-component wave functions that describe either left- or right-handed waves. Suddenly he realized what was going on in the weak interaction: only the left-handed waves play a role in the weak vertices!

In concrete terms, this means that the weak interaction is all the more anxious to create a left-handed particle in a decay, the closer its speed comes to the speed of light. A massless neutrino is always produced left-handed, because it moves at the speed of light. The relatively light electrons also usually move quite fast after being created, e.g., in beta decay and are therefore also preferably produced left-handed, if possible. In contrast, muons are around two hundred times heavier, and the weak interaction is more likely to allow them to be right-handed, as they usually move much more slowly than electrons.

All this came out without difficulty in Feynman's theory. Feynman did all kinds of calculations overnight and was able to reproduce the results of Lee and Yang. However, when he calculated the beta decay of the neutron, the agreement with the known data was less good. The situation was rather confusing – was his theory wrong, had he miscalculated, or were the data unreliable?

If only Feynman had been more persistent and had taken a closer look at the quality of the beta decay data. He had indeed been on the right track, but he had let himself get confused. Therefore, he had not pursued the matter further and left the USA for Brazil, where he spent the summer.

Back at Caltech, he talked to several physicists, including Hans Jensen, Aaldert Wapstra, and Felix Boehm, who informed him of their results on the beta decay of the neutron. These fit much better with Feynman's theory, but contradicted what many other physicists thought was right. Much of the beta decay data seemed questionable, and Feynman was told that his young colleague Murray Gell-Mann was of the opinion that the weak interaction might even be *V minus A* instead of *S* or *T*.

These were, of course, mathematical expressions that only the expert could understand. So Feynman knew immediately what Gell-Mann was saying. He jumped up abruptly and shouted: "Then I understand EVERYTHING!" *V minus A* was nothing other than a mathematical description for the fact that only the left-handed waves at the vertices play a role (*V* stands for *vector* and *A* for *axial vector*, but we don't have to worry about that here). The older data seemed to have indicated an *S* (*scalar*) or *T* (*tensor*) for the weak interaction. These were mathematical descriptions of different spin couplings at the vertices and did not fit Feynman's theory. They did not fit with reality either. But if *V minus A* was the right structure, there was a good chance that Feynman's theory was correct and would also describe the beta decay of the neutron well.

Feynman got very excited and once again threw himself into a night full of calculations. He had to make sure his theory worked. First, he calculated the lifetimes of the muon and the neutron, which were related if his theory was correct. His calculation corresponded to the experimental result with a discrepancy of only 9%. That wasn't bad!

Many other things that Feynman calculated were also in good agreement with the data, and he became more and more excited. This was the first and only time in his career that he knew a law of nature that nobody else knew – or so he thought. He had always hoped for this – to be the first to find a fundamental law of nature, just like his hero Paul Dirac, who had found the Dirac equation. Until then Feynman had, in his view, only applied the theories of others, found new methods for their calculation as in QED, or explained physical phenomena using the well-known Schrödinger equation, as he had done for superfluid helium. But now he was on the trail of something fundamentally new for the first time! He knew how the weak interaction worked and why it violated mirror symmetry. And indeed, it did so in the most extreme way imaginable, allowing only left-handed waves at the vertices.

Feynman called his sister Joan and thanked her – without her he probably wouldn't have delved so deeply into Lee and Yang's paper. While he had felt uncomfortable and out of touch before, he was now back at the forefront of physics again. He fell into a true frenzy and did all sorts of calculations during the night. Everything was just right. Only the discrepancy of 9% in the relationship between the lifetimes of the muon and neutron bothered him somewhat.

The next morning Feynman ran to the experimental physicists at Caltech and told them about his results. When he mentioned the 9% difference, they drew his attention to the fact that there were new readings on lifetimes. The old measured values had apparently been corrected by 7% – only in which direction? Had the difference between Feynman's calculation and the experiments increased to 16% or decreased to 2%? They looked closely at the data in the publications and Feynman breathed a sigh of relief: it was only 2%. Thus his theory agreed well with the measured data, in fact, to within experimental error!

So Feynman's theory was correct – or at least almost correct. A small correction factor still needed to be added, as Nicola Cabibbo found in 1963. This correction factor was related to the fact that up and down quarks like those in the neutron or proton are not the only ones that exist in nature. There are also heavier unstable quarks like the strange quark mentioned earlier. The strength of the vertex between the up and down quarks must therefore be slightly reduced in the theory of weak interaction in order to allow a vertex between up and strange quarks, which causes the decay of the K^+, for example. With this correction, the theory corresponds perfectly to experiment.

Feynman was mistaken on one point though, as he soon came to realize: he was by no means the only one who had found the correct structure of the weak interaction. However, that didn't spoil his fun, because he had solved the problem completely on his own. That was exactly what was important to him. He had proven to himself that he was able to do it, and he had no problem with the fact that others had also succeeded.

If Feynman had not travelled to Brazil in the summer of 1957 and had instead pursued his idea immediately after the conference, he could actually have been the first. In the meantime, however, Robert Marshak – one of the co-founders of the Rochester Conference – and his student George Sudarshan had also come across the mathematical *V minus A* structure of the weak interaction in the summer of 1957. However, they failed to publish their idea early enough, something they both regretted very much later on.

Feynman and Gell-Mann: An Unmatched Pair

Murray Gell-Mann (Fig. 4.20), who had introduced the term *strangeness* a few years earlier, was now also working intensively on the problem of parity violation and had come to the same conclusions as Feynman, Marshak, and Sudarshan.

Gell-Mann had come to Caltech about two years earlier to work with Feynman – his office was almost right next to Feynman's. Although

Fig. 4.20 Murray Gell-Mann (born 1929) in 2007 (© I, Joi; https://commons.wikimedia.org/wiki/File:MurrayGellMannJI1.jpg; CC-BY-SA 3.0 Unported)

Gell-Mann was only 27 years old in 1957 – eleven years younger than Feynman – he had no problem keeping up with Feynman in physics. He was even in his very best years, while Feynman had already passed the peak of his creative power.

At first, a promising collaboration developed, in which Dick and Murray, as many called them, roamed the Caltech campus together and got involved in intense discussions that could last for hours. Over time, however, this collaboration increasingly turned into rivalry – the personalities of these two great physicists were just too different. Each respected the other's abilities in the field of physics, but in many other things they had very different attitudes. Where Feynman was casual, Gell-Mann was someone between correct and pedantic. While Feynman came along in shirtsleeves and didn't care much about formalities and etiquette, Gell-Mann was cultured, always well dressed, and paid much attention to the right pronunciation of words from foreign languages. Feynman, on the other hand, spoke more like a New York taxi driver.

You can get a good impression of their relationship if you watch the interview entitled *Murray Gell-Mann talks about Richard Feynman* on YouTube. There Gell-Mann describes the collaboration with Feynman as very exciting and pleasant in the first years. They constantly surprised each other with new ideas and then got involved in those intense discussions. "It was quite fun", says Gell-Mann with a smile. But after a while the fact that Feynman was constantly occupied with self-portrayal and creating a certain image of himself finally got on his nerves. According to Gell-Mann, he was constantly trying to create anecdotes about himself and always wanted to be different from others, an independent thinker, not an ordinary person. In Gell-Mann's view, the whole thing was sometimes close to ridiculous.

You can tell from these words that it had not been easy for Gell-Mann to work with Feynman in the long run. Many of Feynman's doctoral students also came to realize this. It should not be forgotten that Feynman was rather a loner in physics, despite his charismatic personality – he just loved too much to get behind the secrets of nature himself.

Gell-Mann was probably also a little jealous of the attention Feynman attracted as a gifted speaker and storyteller. In physics, Gell-Mann was as good as Feynman, and when it came to general education, hardly anyone could match Gell-Mann. "The Man Who Knows Everything" titled the New York Times in 1994 in an article about Gell-Mann which went on: "The essence of Murray Gell-Mann is to know."

Yet Feynman was always the more famous of the two. For example, Gellmann's book *The Quark and the Jaguar* could never match Feynman's books in popularity – amusing anecdotes were simply not Gell-Mann's strength. While Feynman thought in pictures rather as Einstein had done and could explain the secrets of physics very vividly, Gell-Mann's world of ideas was more abstract and could not be communicated so easily to a broad audience.

Their rivalry became clear for the first time when Gell-Mann went back to the Caltech shortly after Feynman's return from Brazil in 1957. There he realized that Feynman had also discovered the correct structure of the weak interaction and was about to publish it. That was exactly what Gell-Mann also had in mind.

In order to curb the emerging competition, the head of the physics department at Caltech, Robert Bacher, urged Feynman and Gell-Mann to publish their ideas in a joint paper. They actually agreed and submitted their six-page text entitled *Theory of the Fermi Interaction* to Physical Review in September 1957, where it appeared in January 1958 – a milestone in theoretical physics. It may have comforted Marshak and Sudarshan somewhat that Gell-Mann at least thanked them at the end of the article for the valuable discussions he had had with them previously.

As it turned out, Feynman and Gell-Mann's theory worked very well and could explain all possible weak decay processes. Soon there was no doubt that this was also true for the beta decay. How could anyone have thought that the beta decay data said otherwise? It was precisely this that had confused Feynman so much at the beginning that he had not initially pursued his idea further.

Feynman decided to get to the bottom of the matter and look at the old data in the original publication. When he had picked out the relevant paper, he was shocked. He remembered that he had seen the article before and, when looking at the data, had thought: "That doesn't prove anything!" The data had been far too inaccurate to draw any conclusions about beta decay. And Feynman had actually realised this! It was just that he had forgotten it again and instead relied on the opinions of the other experts. That was never to happen to him again, he decided. In the future he would no longer trust any expert opinion and always check everything himself – and that's exactly what he did.

5

From Researcher to Teacher and Nobel Prizewinner

In 1957 a dream had come true for Feynman. He had discovered a new law of nature using his own methods: the left-handed nature of the weak interaction. It had been a pure coincidence that other physicists had also made the same discovery by their own means, as he later found out.

In the following years, Feynman continued to be interested in many topics of physics and provided some innovative ideas, for example in nanotechnology and quantum gravity. However, when it came to consistently pursuing these ideas, he tended to stay on the sidelines and leave the field to others. Perhaps it was simply too time-consuming for him to work out his thoughts down to the last detail, because in addition to research, the teaching of physics to students and indeed to a broader public was becoming increasingly important. The famous researcher became an equally famous teacher, whose talks and lectures set new standards and still continue to evoke the enthusiasm that Feynman himself felt.

One of the highlights of his scientific career came when he was awarded the Nobel Prize in Physics in 1965, along with Schwinger and Tomonaga "for their fundamental work in quantum electrodynamics, with deep-ploughing consequences for the physics of elementary particles". At first Feynman was ambivalent about the Award Ceremony, but he finally came to terms with it and gave one of the best Nobel Prize speeches ever made.

Feynman's interests were not limited to the natural sciences at that time: he further perfected his beloved bongo playing and discovered a passion for painting. However, the most important change occurred in Feynman's private life: Feynman would finally find the woman of his dreams and thus fill the gap left by the tragic death of Arline.

J. Resag, *Feynman and His Physics*, Springer Biographies,
https://doi.org/10.1007/978-3-319-96836-0_5

5.1 Marriage, Family, and the Nobel Prize

After Feynman had separated from his second wife Mary Lou in 1956, he was a free man again. Just as before this unhappy marriage, he plunged once again into various affairs and amours, but these did not go too well and sometimes even got him into trouble. Moreover, he did not find in them what he was ultimately looking for: a rich relationship of the kind he had had with Arline more than a decade earlier.

In the summer of 1958, Feynman – now forty years old – went off to the annual Rochester Conference, which for the first time was not held in Rochester, but in Geneva.

Gweneth – The Woman of a Lifetime

On the beach of Lake Geneva, he happened to meet a young woman wearing a blue bikini and got into conversation with her. Her name was Gweneth Howarth. She came from a small village in Yorkshire, England, and despite being only 24 years old, she had decided to explore the world on her own and travel all the way to Australia. So she had bought a ticket to Geneva, where she now worked as an au pair to make enough money to continue her journey.

Feynman admired Gweneth's courage and adventurous spirit – had he possibly found a soul mate in her? He found her so sympathetic that he invited her to accompany him to California and work there as a housekeeper for him. But Gweneth's destination was actually Australia, so at first she refused his offer. On the other hand, she got on very well with Feynman, and promised to think about it. Two months later she finally agreed.

However, it was not easy to overcome the bureaucratic obstacles to her immigration to the USA. After all, a 40-year-old man was trying to bring a young woman to the States to live with him in the same house and they were not even married – this seemed morally questionable in the prudish 1950s and may well even have been illegal. What was more, someone had to be available as guarantor by law to support Gweneth financially if necessary. Luckily, the experimental physicist Matthew Sands, a friend of Feynman's, offered to take this responsibility, although Feynman would, of course, have stepped in financially if necessary. So Gweneth was finally brought to the USA in the summer of 1959, now 25 years old.

Gweneth lived in a room at the back of the house, while Feynman lived at the front. Despite Feynman's tendency to engage in quick affairs, Gweneth's role was initially exactly the one for which he had brought her to the USA: she was his housekeeper. This was apparently a role that urgently needed to be filled in Feynman's bachelor household. He had optimized his wardrobe to include several identical pairs of shoes and always the same dark blue suits with white shirts. Besides that, he didn't have either TV or radio.

Gradually, the nature of their relationship must have changed. They simply made a good match despite the age difference of sixteen years, and Feynman was very happy to have Gweneth with him. Finally, after careful consideration, he asked her to marry him in the spring of 1960.

This proposal came as a surprise to Gweneth, so she asked him for some time to consider it. But the next morning she had already made up her mind and accepted Richard's proposal. They were married on September 24, 1960.

But could this relationship really work? After all, Feynman had had little luck with women in all the years since Arline's death and his second marriage with Mary Louise Bell had failed miserably. But Gweneth was very different from Mary Lou: she was as courageous and adventurous as Richard, and at the same time she was much more loving and sociable than the rather cool and distant Mary Lou. Feynman also became much more balanced – it had not always been easy to get along with him before that. Finally he seemed to have found a woman who could fill the huge gap Arline had left in his life.

Just like Murray Gell-Mann, who had also married a woman from England, Richard and Gweneth bought a house in the hills of Altadena to the north of the university – the two houses were not far apart. In the early days Murray and his wife Margaret came to visit several times before their friendship cooled off and the feeling of rivalry between the two men finally began to prevail.

The friendship and love between Richard and Gweneth, on the other hand, lasted for a lifetime – they remained happily married until their deaths. Two years after the wedding their son Carl was born, and in 1968 they adopted a little girl: Michelle. Feynman was happy – he had always wanted children. In later years he would show his children all the miracles of nature, just as his father had done with him. Carl proved to be very curious and interested in everything that his father explained to him. Michelle, on the other hand, was less interested in those things. When he told her the Sun was made of gas, she said she didn't like gas. But she loved his jokes, for example, when she was allowed to twist his nose like the knob on a radio and he imitated different radio stations. Michelle was just different from Carl, and Feynman loved them both.

So Feynman finally found happiness in his private life and proved to be a loving father and husband. Forgotten were the times when he restlessly threw himself from one affair after another without finding what he was looking for. In addition, Richard's mother Lucille had already moved from New York to Pasadena in 1959 to be close to her son – their quarrel following his marriage to Arline many years before had finally been settled.

Nobel Prize in Physics 1965: Feynman, Schwinger, Tomonaga

In addition to his private fortune, an event came up in 1965 which for most scientists would probably be the absolute culmination of their scientific career: Feynman was awarded the Nobel Prize in Physics, together with Julian Schwinger and Shin'ichirō Tomonaga. Notably, Freeman Dyson wasn't one of the prizewinners, which is a great shame, because he certainly deserved it.

When Feynman was woken up by a reporter at four o'clock in the morning and told that he was to be awarded the Nobel Prize, he did not really know at first whether he should be happy about it. He disliked pompous events and the award ceremony in the presence of the Swedish royal family certainly fell into this category. He also wondered how relevant it was that after many years a committee in faraway Sweden now considered his work worthy of the famous prize. Wasn't it enough reward to feel the satisfaction of solving the difficult problems of QED on his own? "The prize is the pleasure of finding things out" – that was his motto.

On the other hand, Feynman also felt flattered at being invited to join the community of the great Nobel Prize winners such as Max Planck, Albert Einstein, Niels Bohr, and Paul Dirac. Despite his other concerns, it was a great honor to be awarded the Nobel Prize. Moreover, rejecting the Nobel Prize would have caused a great fuss and might have made him appear arrogant or conceited. So after some hesitation Feynman finally accepted and travelled somewhat nervously to the award ceremony in Stockholm.

It was very fortunate for us that Feynman did accept the prize because he gave a most extraordinary Nobel Prize speech in Stockholm. He provided his listeners with a deep insight into the physical world as he viewed it, and explained in detail how his ideas about QED gradually developed. It is a real pleasure to read the speech – and it can be found on the Internet at www.nobelprize.org. Instead of bothering his listeners with mathematical details, as Schwinger did to some extent in his speech, Feynman describes in his usual entertaining way how he and Wheeler had tried to eliminate the infinity of self-energy from electrodynamics, how he had found the path integral formulation of quantum mechanics thanks to Dirac, and how he had finally developed his intuitive method of Feynman diagrams, with which so many things in QED could suddenly be calculated much more easily and simply than before. The speech is written so vividly that we almost get the feeling of participating directly in Feynman's thought processes. Feynman's ability to captivate his audience is visible in all its brilliance. As was so often the case with him, we have the feeling of really understanding the complex subject matter.

All in all, Feynman enjoyed his trip to Sweden. And since he was now in Europe anyway, he travelled to Switzerland to visit the European center for nuclear research CERN near Geneva – the very city where he had met Gweneth seven years before. During his lecture there, he appeared in the fine suit he had worn to the King's Dinner and announced that the Nobel Prize had changed him – from now on he loved this suit. Loud protests arose – Feynman had never worn such suits in his lectures before. Thereupon he took his jacket off, loosened his tie, and thanked his listeners for bringing him back to the real world. The royal pomp was gone and Feynman was himself again – amusing, brilliant, and down-to-earth.

5.2 Nanotechnology: There's Plenty of Room at the Bottom

In December 1959, Feynman gave a remarkable lecture at the annual meeting of the American Physical Society at Caltech. His vision in this talk anticipated the development of a field of research and an industry that began to develop its potential only a few years ago: nanotechnology. The title of his presentation was *There's Plenty of Room at the Bottom*.

An Encyclopedia on the Head of a Pin

What would happen if things could be manipulated and controlled on a tiny scale? With this Feynman did not mean, for example, the scale of microfilms, in which letters were reduced by a factor of 20 to 50 – something like this was already possible at that time. When he spoke of the bottom he meant a much smaller scale. For example, was it possible to write all the volumes of the *Encyclopaedia Britannica* on the head of a pin?

Fig. 5.1 If you reduce the size of the letters by 25,000 times, you can write the complete *Encyclopaedia Britannica* on a single pinhead

To achieve this, the font size of the encyclopaedia must be reduced by a factor of 25,000, as Feynman shows – around a thousand times smaller than the scale used to produce microfilms (Fig. 5.1). A full stop at the end of a sentence would then have a diameter of about 8 nm, where one nanometer is a billionth (10^{-9}) of a meter. For this reason, such extreme miniaturization is called nanotechnology.

A typical metal atom on the pinhead has a diameter of about 0.25 nm, so the whole full stop would have a diameter of just 32 atoms. Such a full stop would no longer be visible to the naked eye, because the human eye has a resolution of only about 0.2 mm. Even with a standard light microscope, the full stop would not be visible, because the wavelength of visible light, which is a few hundred nanometers, limits the resolution of optical instruments and is significantly larger than our full stop. The nanometer range is ten to a hundred times below the scale that we can see with light.

On the other hand, despite its small size, the area of the full stop would still contain about 800 atoms, which should not be a physical problem, as Feynman states: "There is no question that there is enough room on the head of a pin to put all of the Encyclopaedia Britannica."

Microscopes for the Nanoworld

But would we be able to read such tiny letters on a pinhead? Feynman also investigated this question and concluded that it should be relatively easy with an electron microscope – even in the late 1950s! An electron microscope does not work with light, but with a beam of electrons. According to the rules of quantum mechanics, these particles behave like a wave with a much shorter wavelength than light. This is sufficient to clearly identify the tiny letters on the pinhead!

At the time, the electron microscope seemed to Feynman to be the ideal tool for the nanometer scale, and in his talk he called for further improvements to its resolution. In this way he hoped that even single atoms would eventually become clearly visible. Many problems in chemistry or biology could thus be solved simply by looking directly at the atomic structure of the molecules.

Over time, it has become clear that the electron microscope is not necessarily the ideal tool for the nanoworld. If, for example, an object is to be illuminated with electron beams, it must first be embedded e.g., in a synthetic resin mass, and then sliced into extremely thin slivers so that an electron beam can penetrate it. An entire biological cell cannot be penetrated by an electron beam anyway, and the cell would hardly survive such a process intact – the high-energy electrons would cause too much damage. However, considerable progress has been made in recent years, especially in the field of so-called cryo-electron microscopy, and electron microscopes can now be used to decipher, e.g., the structure of large protein molecules – and indeed, in 2017 the Nobel Prize in Chemistry was awarded for this ingenious method.

Instead of illuminating an object, it can also be coated thinly with metal and then scanned point by point in a scanning electron microscope under high vacuum with a finely focused electron beam. This beam knocks out other electrons from the metal surface, and it is then possible to compute an image of the surface (see Fig. 5.2). But no matter how it is used, electron microscopy is always a technically complex matter in which samples have to be prepared with considerable effort and can then only be examined with the electron beam in a good vacuum. Feynman's vision of exploring the nanoworld with the electron microscope has therefore only been partly realised.

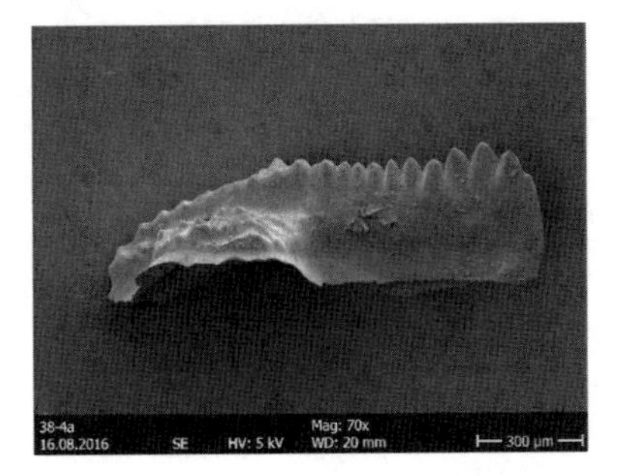

Fig. 5.2 Side view of a conodont about the size of a grain of sand, taken with a scanning electron microscope. Conodonts are tiny tooth-like microfossils and probably originate from small primitive fish that populated the oceans 540–200 million years ago. They can be found in large numbers and in a wide variety of forms in sediments from this period. Even the finest structures can be detected with the scanning electron microscope. The conodont shown here comes from the area of Arnsberg (Sauerland, Germany) and is about 340 million years old. Credit: Kevin Resag

Fig. 5.3 A 40-nanometer-wide NIST (*National Institute of Standards and Technology*) logo made with cobalt atoms on a copper surface. *Credit* Joseph Stroscio; Robert Celotta/NIST; https://commons.wikimedia.org/wiki/File:NIST_HipHopAtomLogo.jpg, http://www.nist.gov/public_affairs/releases/hiphopatoms.cfm

In 1981 – more than twenty years after Feynman's visionary lecture – the German physicist Gerd Binnig and his Swiss colleague Heinrich Rohrer developed a new tool in the IBM research laboratory near Zürich that has since proven to be extremely useful in nanotechnology: the scanning tunneling microscope. With a very fine needle tip, this microscope can scan a surface extremely accurately and even almost sense individual atoms. The tip is controlled by a quantum mechanical tunnelling current between tip and surface in such a way that the distance from the tip to the surface always remains constant. The microscope thus determines an elevation profile of the sample surface, which can then be displayed by a computer in three dimensions.

With the scanning tunneling microscope and similar tools, surfaces can not only be scanned with atomic resolution, but individual atoms can be precisely manipulated. This was first achieved in 1989, when IBM scientists arranged 35 individual xenon atoms on a frozen nickel surface in such a way that they formed the IBM logo – an example that has been repeatedly copied since then (Fig. 5.3).

Hands Control Hands

Feynman could hardly have foreseen such possibilities in 1959. In his lecture, he had come up with another, highly original method which he hoped could be used to manipulate matter on the nanoscale. In nuclear power plants, there are mechanical hands that can be moved in a similar way to real hands via a remote control system. These "slave" hands imitate the movements of the controlling "master" hands and move synchronously with them. This makes it possible to handle hazardous radioactive materials without touching them directly.

Now the mechanical hands do not necessarily have to be the same size as human hands. They could also be significantly smaller so that they can be used on a millimeter scale, for example. These millimeter-sized hands could now control other hands, which are significantly smaller again and so on – until we reach the nanoscale.

Would a cascade of ever smaller mechanical hands work? That is difficult to say, but it is clear that at least some serious problems would have to be solved. On the nanoscale, physics is very different from our familiar environment. When an object is reduced in size, its surface area decreases more slowly than its volume, so that everything related to surfaces or cross-sections – for example frictional forces or adhesion to surfaces – becomes very important in the nanoworld. For a bacterium, clear water is a viscous liquid which acts like honey: without propulsion it immediately gets stuck in it.

In addition, the permanent thermal motion of the atoms causes an object to vibrate more and more or to bounce back and forth randomly the smaller it is. On the lowest level, the mechanical hands will start to tremble, which will make it difficult to work accurately. In addition, the rather weak van der Waals forces between the atoms will make all sorts of things stick to the tiny nano-hands.

Nanomachines

Can there be any moving machines in the nanoworld at all? The answer is clearly: YES! Our cells are an impressive example of how a complex network of nanomachines can work together and make life possible. However, these cellular nanomachines are not made of metal, but of organic macromolecules, and they also work very differently from the machines we are used to in our macroscopic world. The unavoidable thermal motion of the atoms does not harm them, but ensures that randomness will always bring appropriate parts together and set them in motion, acquires energetic molecules and raw materials, and removes finished products. Electric charge differences in the atomic structure of the molecules ensure that the correct parts come together. At the same time, the omnipresent water component in the cell fluid shields the van der Waals forces to such an extent that the molecules do not stick together without specific reason. Water even helps to configure the molecules, because they try to shape themselves in such a way that hydrophilic – i.e., water-loving – areas turn outwards towards the water, while hydrophobic (water-repellent) areas turn inwards to hide from the water.

An example of such a nanomachine is the flagellum on certain bacteria. A tiny nanomotor rotates a flagellar filament so that it drives the bacterium through the viscous water like a ship's propeller. The motor, which consists of special protein molecules, is located in the outer cell membrane of the bacterium and is driven by a gradient in proton concentration between the interior and exterior of the cell. At around 15–20 nm, the filament is only about twice as thick as the full stop in Feynman's encyclopedia on the pinhead.

Living cells contain many more of these nanomachines, which perform tasks of chemical synthesis, transport molecules through the cell or pass them through cell membranes, read and process genetic information, and much more. When many of these nanomachines come together and pull in the same direction, they can even bring about macroscopic movements – our muscles are the best example.

Today, chemists are already able to produce a whole range of different molecules that can function as nanomachines in many different ways. There are nanoswitches, motors, propellers, rings, rods, and gears. In 2016, the three chemists Jean-Pierre Sauvage, Fraser Stoddart, and Bernard Feringa received the Nobel Prize in Chemistry for the development of such molecular machines. But the whole field is still in its infancy. It is not easy to find suitable applications for all these components and to implement them in a useful way. However, the potential is enormous, and sooner or later the first nanomachines will appear in our everyday lives.

In 1959, when Feynman gave his talk, all this was still a utopian dream. Accordingly, his lecture is a colorful mixture of forward-looking ideas, little calculations, bold speculations, and of course a few misjudgements.

A good example of the latter concerns chemical synthesis. Feynman says that a chemist does mysterious things when he wants to produce a molecule: he mixes this and that, shakes it, and fiddles around. Only at the end of a difficult process like this does he usually succeed in synthesizing what he wants. A physicist, on the other hand, would have to be able to produce any chemical substance that a chemist writes down – simply by placing the atoms where the chemist wants them.

Well, it's probably not that simple after all. On the other hand, there is also something true about this: For example, when a protein strand is synthesized, a cell uses the DNA molecule to control where the various amino acids are to be placed in the protein molecule. So there are indeed possibilities for arranging individual building blocks within a molecule, even if these building blocks comprise more than just one atom.

Fast Computers are Tiny

Particularly interesting are Feynman's remarks about storing information in a very small space. We have already seen above that all 24 volumes of the Encyclopaedia Britannica would fit on one pinhead. If we use not only the surface but also the inside of the material for storage, we can store a lot more information in it. Feynman calculates what would happen if each bit of information could be represented by a cube of 5 by 5 by 5 atoms – that is, by about 100 atoms. Then all the books in the world could be stored in a cube with an edge length of only 0.1 mm. A tiny piece of dust that can hardly be made out by the human eye would offer enough space to contain everything human beings have ever known! "Don't tell me about microfilm!" joked Feynman. There's really plenty of room at the bottom!

Life has been using this opportunity for a long time, as Feynman pointed out – he was very interested in biology at the time and even took part in some research projects. One of the reasons for his interest was certainly a great discovery that had taken place just six years earlier, and in which his Caltech colleague Max Delbrück was also involved. In 1953 James Watson and Francis Crick had discovered the double-helix structure of DNA. Since then it had become clear that all our genetic information – what we look like, how our anatomy is organized, and how our cells function – is stored in the DNA molecules in each individual cell. Only about 50 atoms are needed for one bit of this information. Such a density of information is thus quite feasible and does not contradict the laws of physics.

Such storage densities would be very interesting for computers! The only problem is to be able to read and write the information. And to be able to process it quickly, whence the need for fast computers. In Feynman's day, computers still filled entire rooms. Such computers could not be particularly fast, because information cannot move from one component to the next faster than the speed of light. Increasing the number of circuits in a computer to increase its performance would have made computers of that type too large to work effectively.

Fast, powerful computers have to be small, and Feynman made a strong plea in favour of this in his presentation. There was a lot of room to make them smaller, and he could see nothing in the laws of physics that would disallow this. It seemed possible to use wires with a diameter of only 10–100 atoms and transistors that were only a few tens of nanometers wide.

Such computers could become so powerful that they would have completely new qualitative features. Feynman surmised that they might be able to draw their own conclusions and gain experience about what was the best solution to a problem. They would be able to recognize faces in photos – an ability that we humans can easily master with our countless microscopic connections between the neurons of the brain.

In the meantime, we have actually made good progress in this direction, in both the production and the application of computers! Today, any smartphone has more computing power than the room-filling computers of those days. Automatic face recognition is no longer a big problem for computers, and understanding spoken language is also becoming more and more common in our everyday lives – just think of speech recognition software like Siri or Cortana. Computers can learn to drive and they play chess so well that even a world champion hardly stands a chance. Even in the Asian game Go, which is particularly difficult for computers to calculate, these machines are beginning to gain the upper hand. In March 2016, for example, AlphaGo beat one of the world's best Go players, South Korean Lee Sedol, in five games four to one. These are areas where Feynman's vision has become a reality.

If computer chips are reduced even further down to the atomic scale, new effects will come into play that do not yet play a role in today's computers. In a circuit consisting of a few atoms, a lot of new things will happen that represent completely new opportunities for design, wrote Feynman. At this level, the rules of quantum mechanics would apply! At that time, the idea of a computer using quantized atoms appeared to be a very long way in the future, but this has now become a very active field of research. In theory, quantum computers can do many things that a normal computer cannot. The difficulty is to shield the computing atoms so well from their environment that their quantum states are not permanently destroyed again – there is still a lot of work to do here!

Feynman's Competition for Nanotechnology

Feynman ended his presentation with the announcement of two prizes:

- $1000 to the first person who could take the information on the page of a book and put it on an area 1/25,000 smaller in linear scale in such a way that it could be read by an electron microscope.
- $1000 to the first person who could make a working electric motor that could be controlled from the outside and was only the size of a cube of side 0.04 mm.

He did not expect it to take long before these prizes were claimed. And indeed, just a year after Feynman's talk, William McLellan, an American electrical engineer, presented a tiny engine that met Feynman's specifications. However, Feynman was a little disappointed, because no new micro- or even nanotechnologies had been necessary for the construction of the motor – the techniques available at that time were already sufficient. So the first $1000 became due rather soon, and Feynman, who had just returned from his honeymoon, had to explain to his young wife Gweneth why he was giving away $1000.

The other task proved to be much more difficult. After all, reduction by a factor of 25,000 means that the complete *Encyclopaedia Britannica* can be written on a pinhead. It was not until 1985 – more than 25 years later – that Stanford graduate student Tom Newman succeeded in putting this into practice: he wrote the first page of Charles Dickens' historical novel *A Tale of Two Cities* in the desired size on a pinhead using an electron beam. Later on it was not at all easy to find the tiny text on the comparatively huge pinhead.

Two years before this masterpiece – in 1983 – Feynman gave another lecture entitled *Infinitesimal Machinery*. There he looked back on what had become of his ideas and predictions in the almost 24 years since his first talk in 1959. For example, he showed around the tiny engine that William McLellan had built, and he presented a picture, reduced by a factor of 30,000, of a book that someone had sent him a few years earlier. So some important progress had indeed been made here.

In this talk Feynman was sceptical about the possibility of tiny nanomachines. Not much had happened in this field. Today we know that it would take much longer for this area to slowly gain momentum. Why were there no nanomachines yet, Feynman asked? There was of course the problem of how such machines could be produced, and Feynman made a few new suggestions. With a smile he also mentioned his earlier idea with the ever smaller remote-controlled hands: "You're right to laugh", he grinned and admitted that this was probably not a sensible technique.

The Rocky Road to Application

Feynman saw a central problem in finding useful applications for nanomachines. For example, one idea would be to punch tiny holes in a thin plate in such a way that they could be opened and closed individually. If we let light shine through these holes, we could create all kinds of light patterns that could also be changed very quickly. The only problem is that a TV can also produce moving images. So where would be the advantage of such a new technology here?

This is exactly the difficulty that nanotechnology still faces today. It is not so easy to find useful applications in which nanotechnology is really superior to established methods. Feynman also admitted that at first he couldn't think of much more than the above example. But then he did come up with some additional ideas.

How about tiny nano rollers on a surface, for example? Then we could very easily wipe away dust and dirt, because it would not stick to the surface thanks to the rollers. This is quite interesting – except that you don't need any rollers at all. Small nano-bumps are already sufficient to reduce contact with the surface to such an extent that dirt is easily carried away by a drop of water. This is exactly how a lotus leaf manages to stay clean over time. Nowadays, this lotus effect is already used for a wide range of products.

Feynman gave some more ideas, such as tiny drills, and so on, and he discussed a nanomachine that could move freely through a liquid like a bacterium. He point out how viscous water is for such a small object, whence we need to think carefully about how to move forward in a viscous liquid.

When we read Feynman's talks, we realize how fascinated he was by the possibilities of the nanoworld. He felt it was similar in some ways to what had motivated Kamerlingh Onnes, whose experimental methods had opened up the field of cryophysics. Here was another apparently bottomless field of investigation, in which the temperature could be lowered ever further. Fundamental new discoveries such as superconductivity and the superfluidity of helium had been the consequences there. What would turn up if we moved into ever smaller dimensions?

If we look at Feynman's first lecture *There's Plenty of Room at the Bottom* from today's perspective, it seems to have been a kind of starting point for the entire field of nanotechnology. His vision that it would be possible to control matter right down through the nanoscale to individual atoms seems in some ways to anticipate current developments in this field.

However, it is questionable whether his lecture really had this significance. When Feynman presented it at the annual meeting of the American Physical Society in Pasadena over fifty years ago, it certainly made a deep impression on the professional audience there. In the following three years, his lecture was also printed several times. But then, for more than twenty years, it disappeared from the scene and was largely ignored. For Feynman himself, his excursion into the nanoworld was probably a minor episode in his scientific career.

Only in the years after 1980 did the development of nanotechnology slowly take off – in 1981, with the scanning tunneling microscope, and in 1985, with the discovery of fullerenes (football-shaped molecules of 60 carbon atoms), for example. At that time some people began to remember the lecture that the now famous physicist had given many years before. The presentation suddenly looked like the founding document for the entire field, providing it with a certain lustre thanks to the glamour associated with Feynman's name.

Presumably, nanotechnology would have emerged even without Feynman's lecture. Nevertheless, it is still a pleasure to read it today and with hindsight to see what has become of the many ideas that Feynman put up for discussion. The lecture was subtitled *An Invitation to Enter a New Field of Physics*. Many scientists and engineers have now accepted this invitation, even though the invitation itself was temporarily forgotten.

5.3 The *Feynman Lectures*

When young people choose to study physics, it is often for one of the following reasons: they have seen Einstein's famous formula $E = m \cdot c^2$ and wondered where it comes from and what it means, or they have heard that gravity curves space and time, that the universe is being driven apart by a mysterious dark energy, or that particles can be both here and somewhere else at the same time in quantum mechanics. These are things everyone really wants to understand! In the words of Johann Wolfgang von Goethe's *Faust*: "That I may understand whatever binds the world's innermost core together (Dass ich erkenne, was die Welt im Innersten zusammenhält)."

The physics we get to know at high school usually has little to offer here. We are introduced to the inclined plane, the motion of a projectile, and Ohm's law. And when we first start studying physics, that doesn't change much. Of course, such basics are important, but it still takes a relatively long time to get to the really interesting topics. What's more, when we study the basics according to the standard textbook, it's difficult to feel the fascination that they once exerted on our ancestors – after all, what is now regarded as an old-fashioned beginner's topic was once at the forefront of research. When Isaac Newton formulated his laws of motion and gravity, which enabled him to calculate the orbits of the planets, the world of the late seventeenth century was deeply fascinated. An ancient mystery had been solved! The world of the heavens obeyed the same physic laws as the world on Earth. Is there no way Newton's enthusiasm for these classical questions could still be conveyed today?

Maintaining the Fascination

At Caltech, this problem was already apparent at the beginning of the 1960s and an attempt was made to encourage the interest of the enthusiastic and smart students coming out of the high schools and into Caltech, as Feynman put it. This included a fundamental revision of the two-year introductory course that all new physics students at Caltech were expected to go through. The aim was to enrich this course with the topics of modern physics and make it more interesting.

However, it was not at all easy to put this ambitious idea into practice, especially as the opinions of the individual professors on the exact design of the course were by no means uniform. Finally, Matthew Sands, who had helped Feynman bring his future wife Gweneth to the USA, had a felicitous idea: he suggested that Feynman should work out the lectures and give them at least once himself. With his enthusiasm for physics, his captivating style, and his didactic skills, he seemed to be just right for the task. The lectures would be recorded on tape, revised, and finally published as a textbook on which the course could be further built up in the coming years.

Feynman gladly accepted the challenge and threw himself into the task at hand with great verve. From 1961 to 1963, he gave his lecture twice a week in front of about 180 students and tried to instill in them the important physical basics as well as providing some insights into modern physics. He did this in a unique way, as only he knew how.

Feynman usually arrived before the students in the large lecture hall and smiled at them as they gradually entered the room. A certain atmosphere of suspense would arise, almost as it would just before a theater performance, until things finally got started. Full of energy, Feynman would walk up and down in front of the large blackboard, his eyes gleaming with pleasure, gesturing and joking, speaking softly and then loudly again, and completely filling the blackboard as the lecture went on (Fig. 5.4). During the two-year course he presented everything worth knowing about mechanics, radiation and heat, electromagnetism and relativity, atoms, gravity, and gases and liquids, and even gave an introduction to quantum mechanics.

Each lecture was filled with original ideas and thoughts, which one hardly found in any other textbook, with deep – sometimes even quite philosophical – remarks and insights, historical notes, hints about practical applications of physical concepts, cross-references to chemistry and biology, and much more. Feynman unfolded all the treasures of his physical knowledge in front of his audience, and tried to teach them, not only facts, but also his own way of thinking.

Fig. 5.4 Richard Feynman giving a lecture at CERN in 1970 (© CERN; https://cds.cern.ch/record/41331)

In the preface he wrote for the printed version, Feynman states that he aimed to address his lectures to the brightest in the class, and to make sure, if possible, that even the most intelligent would be unable to fully grasp everything that was presented. At the same time, however, he also wanted to teach all the students at least a central core or backbone of material in such a way that they could really understand it. He hoped that none of the students would be disconcerted, stressing that he didn't expect them to understand everything, only the main features.

The whole thing was essentially an experiment, and it was not clear whether it would be successful. Feynman himself was rather pessimistic about the level of success when it came to the end of the courses. He did not think that he had done a very good job for the students – the new system was rather a failure, he wrote in his preface to the printed version. Many of his colleagues disagreed and said that some students had come along very well and had shown great interest. Well, it was exactly these students that Feynman had been trying to get at, and he had apparently succeeded in doing so. But was this any great achievement, he asked himself, and quoted the British historian Edward Gibbon: "The power of instruction is seldom of much efficacy except in those happy dispositions where it is almost superfluous."

Had Feynman perhaps left many students behind with his ambitious lectures, something he had really wanted to avoid? In his preface to the *New Millennium Edition of the Feynman Lectures on Physics* in October 2010, the American theoretical physicist and Nobel laureate Kip Thorne writes that he asked seventeen former students from Feynman's 1961–63 class about their experiences at the time. Most of them had found the lectures to be one of the highlights of their college years. It was like going to church, like having a transformational experience. Some said they had never missed a lecture and could still feel Feynman's sheer joy before scientific discovery. Feynman's lectures had had an emotional impact that was difficult to capture in the later printed version.

However, some of the students interviewed also had negative memories. The lectures alone were not sufficient to solve the exercises given later. Feynman's often artful thought processes were sometimes difficult for the average student to understand, because most students did not yet possess the intuition and experience of a trailblazing physicist. Listening to the lecture, the presented material seemed exciting and understandable, but if someone wanted to understand it in detail later, the transcript they had made could easily look like unreadable Sanskrit. Unfortunately, there was no suitable textbook to match the material either. This problem was only solved a little later by publishing the *Feynman Lectures*, which were produced by Robert B. Leighton and Matthew Sands after considerable editorial effort, using the tape recordings and blackboard pictures from the lectures.

It seems likely that some students in their first year of study were simply overwhelmed by the wealth of material offered, and some of them may have given up in frustration. But there were others who came more and more often: students of higher semesters, doctoral students, and even some members of the physics faculty. For them, the way Feynman presented the subject matter was refreshingly different from the standard method they had got to know in their own studies. Suddenly completely new insights and interrelationships opened up for these listeners, even on topics they thought they knew quite well.

This corresponds to my own experience. When I came across the Feynman Lectures during my physics studies, I was intimidated by the depth in Feynman's presentation of the material there. It would have been simply too time-consuming to work through the extensive texts, especially since as often as not they did not correspond to the standard method used in the lectures at my university. It was more effective to follow the lectures and standard textbooks in order to build up the basic knowledge needed for exercises, tests, and examinations as quickly as possible.

It was only after my studies, when there were no more exams to pass, that I looked at the Feynman Lectures again and was fascinated. With no pressure of time I was now able to follow Feynman's explanations at my own rate, and at the same time I had the background to be able to understand them. Feynman's lectures are probably most useful for those who already have some basic knowledge of physics.

It is most fortunate that Caltech decided from the very beginning to publish the lectures in book form under the title *The Feynman Lectures on Physics* and thus make them available to later generations of physicists. The careful conversion of the lectures into easily readable texts while retaining the typical Feynman style has been outstandingly successful. Of the three volumes, the first deals mainly with mechanics, radiation, and heat, while volume two deals with electromagnetism and the structure of matter, and volume three finally presents the basics of quantum mechanics.

The Feynman Lectures are among the most successful textbooks ever written in physics. They have been translated into over a dozen languages. The English version alone has sold more than 1.5 million copies and is still read by many physics enthusiasts today. Feynman's refreshing way of bringing us closer to physics does not seem outdated in any way even today, and most of his texts read as if they had been written only yesterday. The Millennium Edition of the English version has been freely available on the Internet since 2013 at http://feynmanlectures.caltech.edu/index.html.

One of the reasons why Feynman's lectures are so fascinating is that he repeatedly questions seemingly well-known laws and concepts. What do the laws mean? Why do they look like this and not some other way?

What is Time?

An example of this is the apparently self-evident concept of time. In Sect. 5-2 of the first volume, Feynman asks "What is time?", and continues that it would be nice if we could find a good definition of it. After two unconvincing attempts to define it, he concludes that we should face the fact that time is one of the things we probably cannot define. However, that's not what matters, because it's not how we define time that is important, but how we measure it.

This is exactly the crucial insight needed to understand time – and many other physical quantities – something Albert Einstein emphasized with the following words: "Time is what you read on the watch (Zeit ist das, was man an der Uhr abliest)."

For a watch or clock, one typically uses a periodic event that always recurs in the same way, such as the oscillation of a pendulum in a grandfather clock, a balance wheel in a mechanical clock, a quartz crystal oscillator in a watch, or the resonant oscillation of appropriately prepared atoms in an atomic clock.

Feynman continues by asking how we can be sure that all these clocks are "really" periodic, hence always "ticking" at the same rate. Some omnipotent being could slow down all clocks and all other processes of nature during the night and speed them up during the day. No physical experiment could answer this question. We can only compare different clocks and find out whether their regularities match. "We can just say that we base our definition of time on the repetition of some apparently periodic event", Feynman concludes.

How deep this modern concept of time goes becomes clear when we compare it with the definition given by Isaac Newton, who wrote in 1687:

> Absolute, true and mathematical time, of itself, and from its own nature flows equably without regard to anything external.

So for Newton there was a "true time" with no need for any relationship to an external object such as a clock. His contemporary, the German polymath and philosopher Gottfried Wilhelm Leibniz, took a completely different view, which is much closer to our current understanding of time. For Leibniz, time and motion were closely related – in a world without any change it would be pointless to speak of time at all. Newton's absolute concept of a "true" time is therefore physically meaningless. Feynman clearly elaborates this insight in his lecture! Where else could you find something like that in a textbook for freshmen?

The Laws of Motion

Another example of Feynman's fresh approach to physical foundations is his discussion of Newton's law of motion: *force equals mass times acceleration*, or briefly $F = m \cdot a$. At school or at university this law is often presented without comment so that it can be applied as quickly as possible to calculate the motion of a cannonball or the orbit of a planet, for example. If the aim is to be able to solve exercises and pass exams, this is probably the fastest and most effective way.

Not so with Feynman! In Sect. 12-1 of the first volume he suggests that one ought to stop every once in a while and think about what such a law really means. What is the meaning of force, mass, and acceleration?

If we assume that we know the meaning of position and time – which of course also raises certain questions, as we have just seen – we can define the concept of *acceleration* mathematically, as we did in Sect. 1.3. There we already looked more closely at the concept of *mass* and interpreted it as a measure of the inertia of a body. So let us now concentrate on the concept of *force*. How is this term defined?

Feynman makes the following suggestion: wouldn't it be the most precise and beautiful definition of force simply to say that force is the mass of an object times its acceleration? The formula $F = m \cdot a$ would thus define the force F mathematically. At the same time, however, we also say that this formula is a law of physics. But what does this law mean? With it we would not have said anything new, if the force were simply *defined* as mass times acceleration. Somehow we seem to be moving in circles here!

As Feynman points out, to regard Newton's law of motion merely as a definition of force would make it completely useless, because no prediction whatsoever can be made from a definition, and he illustrates this statement with an example that we already know from Sect. 1.3. Suppose we were to say that an object left to itself keeps its position and does not move. It would only move if a gorce acts on it, and all the more so the larger the gorce – like a train that only moves when an engine pulls it. The word *gorce* is of course supposed to remind us of *force*. We could define gorce simply as the speed of the object and say that everything stands still except when a gorce is acting. This statement would be completely analogous to the above definition of force, and it would not contain any information, because we would only have replaced the word *speed* by the word *gorce*.

So Newton's law cannot be a mere definition. The real content of Newton's law, according to Feynman, is this: the force is supposed to have some independent properties, in addition to the law $F = m \cdot a$, without us being able to say right from the start which exactly these are. So Newton's law is incomplete and says that in a motion we should study the product of mass times acceleration, call it force, and then look for additional properties of this force in specific situations. This is where we should find something if the whole program is to make sense.

If we try the same thing with the concept of gorce, we fail, although the gorce motion law does not look so wrong intuitively. But if we want to describe the motion of a cannonball or the motion of a planet in a physically meaningful way, we quickly reach a contradiction – this is exactly what happened to our ancestors during their unsuccessful attempts over centuries. No conclusive additional properties can be found for the gorce.

With the force, it's different! So Newton could express the force between two celestial bodies by his law of gravity – see next section – and thus calculate the orbits of the planets correctly. This reveals an important property of force: it represents an external influence and thus has a material origin, such as the gravitational attraction of the Sun. This basic *physical* idea goes far beyond a purely mathematical definition of force!

Idealizations – Are Fields Real?

But how can a science like physics work if not everything is defined exactly? Feynman replies that if we insist upon a precise definition of force, we will never get it! Physical terms and laws are not exact. They are idealizations and approximations that can hardly ever be defined precisely. If we look very closely at matter, the concept of force dissolves as soon as quantum mechanics comes into play. A similar thing happens with time – we couldn't define it exactly either; we could only give an instruction for measuring it using clocks. However, if we are dealing with extremely short or very long time intervals, then this measuring rule could reach its limits. It is therefore quite possible that the so-called Planck time (about 10^{-44} s) is the shortest time interval that we can talk about in a physically meaningful way – we will encounter the Planck time again in the next section.

Any simple idea is approximate and every physical concept is an idealization, says Feynman. This attitude can be found in many places throughout the entire lecture series. Another example is the notion of the electromagnetic field. What are these fields really? Feynman addresses this question in the second volume and, not surprisingly, concludes in Sect. 1-5 that this question makes little sense. The only sensible question is: what is the *most convenient* way to look at electrical effects? This is exactly where he had gained so much experience while he was at Princeton, when he and Wheeler had tried to set up a description of electrodynamics without fields and transfer it to quantum mechanics (see Chap. 2 of this book). As Feynman and Wheeler found out, such a description, where charges influence each other directly with nothing in-between, is quite complex. But it is still *possible* to some extent.

So fields are a very useful tool, because they enable a *local* description of electromagnetic phenomena without any direct remote action. We consider that the charges first generate electric and magnetic fields in space, and that it is the values of these fields at the location of another charge that then determine what force acts on this second charge. This kind of description is so useful that we may consider the fields to be *real* in a certain sense.

Indeed, in the nineteenth century intensive efforts were made to define fields as tensions and vortices in a hypothetical medium – called *ether* – that completely fills the whole of space. It was only thanks to Einstein's special theory of relativity that it finally became clear that fields would be better viewed as abstract quantities, without trying to create too concrete an image of them.

Let me give an example. When an electric charge moves uniformly, it generates a magnetic field similar to an electric current. In addition, its electric field lines, which would be spherically symmetrical for a static charge, are spread out ahead and behind the charge and are squeezed together around the sides, as Feynman shows in Sect. 26-2 of the second volume of his lectures (Fig. 5.5 left). But what does an observer see when riding along with the charge? For such observers the charge is stationary, so they would say that there is no magnetic field. Furthermore, the electric field of the charges is now spherically symmetrical, i.e., equally strong in each direction. These comoving observers would thus describe electromagnetic forces on other charges using this static electric field alone. The description of the forces via fields will therefore change depending on the observer!

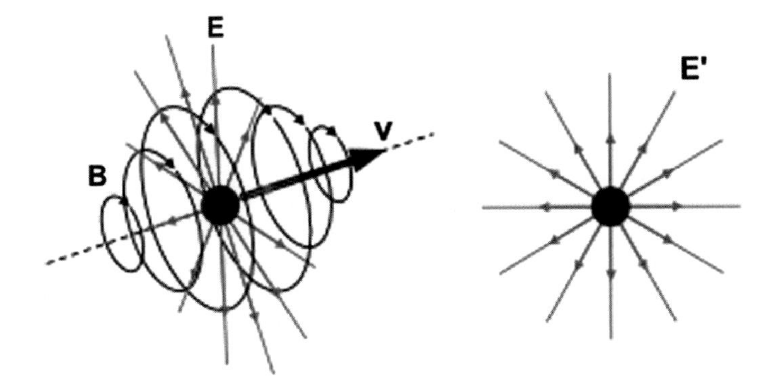

Fig. 5.5 An electric charge moving at velocity v generates an electric field E that is compressed in the direction of motion and a magnetic field B (left). An observer moving along with the charge, however, would describe the effect of the charge on other charges exclusively using a spherically symmetric electric field E', without any magnetic field at all (right)

Strange Quantum Mechanics: The Double-Slit Experiment

An absolute highlight of the Feynman Lectures is the description of the famous double-slit experiment in Chap. 1 of the third volume. Feynman goes straight to the heart of what is strange about quantum mechanics, and what makes it so difficult for us to understand this theory, in a very vivid way. The main feature of the mysterious behavior of quantum mechanics is exhibited here in its strangest form – according to Feynman. So let's take a look at how Feynman describes the experiment.

The experimental setup is very simple. First of all, there is a source that emits a beam of individual particles. It doesn't matter exactly which particles we use – they can be whole atoms, or just single electrons, protons, or neutrons. Photons would also be fine. So let's just say they are electrons.

At some distance from the electron source we set up a wall (usually a thin metal plate) with two small holes in it very close to each other. They can also be thin slits – hence the name *double slit experiment* – but since Feynman speaks of small holes, we shall do the same here. The electrons can fly through these two holes and, some distance behind them, hit a detector screen that registers these incoming electrons. We can imagine, for example, that an electron hitting the screen triggers a small flash of light and leaves a small dot on it.

To carry out the experiment in reality for electrons, extremely small holes or slits must be used, and they must be very close to each other. Therefore, the experiment is usually done with photons – for example laser light – which is much easier. But electrons are also possible, as Claus Jönsson from Tübingen first showed in 1959 (Zeitschrift für Physik A, Vol. 161, No. 4). Feynman probably didn't know anything about this at the time of his lecture – at least he spoke against trying to set this experiment up in reality, because the apparatus would have to be made on an impossibly small scale. As with the tiny electric motor in the previous section, Feynman apparently underestimated the abilities of his colleagues in experimental physics and engineering.

Which pattern of points do we get on the detector screen as time goes by? Where do the electrons hit the screen? Suppose the electrons behaved like macroscopic objects, such as bullets. Then they would have to fly through either one hole or the other. Since the holes are quite small, many electrons would also be deflected at the inner edges of the holes, so that their path there would make a kink. In the end, a relatively fuzzy image of the two

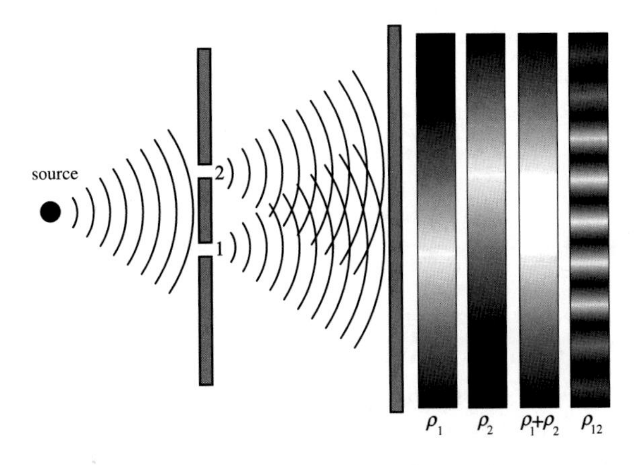

Fig. 5.6 Double-slit experiment with electrons. The four bars on the right represent the probability of electron hits on the detector screen during different test runs – brighter regions correspond to more hits. P_1, P_2, and P_{12} are the probabilities if only hole one is open, only hole two is open, or both holes are open. The bar showing $P_1 + P_2$ would be the expected result for two open holes if electrons behaved like bullets. Instead, a pattern of stripes is found in the experiment (right bar)

holes should form on the detector screen due to the incoming electrons. If we leave only one hole open, we should only get the fuzzy image of the open hole, of course.

Now let us look at the real experiment to see whether this expectation is confirmed. If we leave only one of the two holes open, the above prediction turns out to be correct: we see a fuzzy image of the open hole in the dot pattern of the incoming electrons on the screen, apart from a few details that are not important here. But if we leave both holes open, something completely unexpected happens: the pattern that develops here on the screen over time looks completely different from the sum of the two patterns with only one open hole. Instead, we see a pattern of stripes alternating between very many and very few electron hits (Fig. 5.6). The chances of hitting a certain area on the screen through hole one alone and hole two alone do not simply add up to the chance of hitting that area when both holes are open at the same time.

There's no way to explain this with bullets. Electrons obviously behave in a different way – but how? The behavior of waves gives us a clue. Waves can interfere, e.g., a wave trough and a wave crest can cancel each other. If waves are sent through two adjacent openings, alternating stripes with "constructive" and "destructive" interference, i.e., high and low wave intensity, are formed behind them.

That's exactly what we need. If both holes are open, the wave can be less intense in some places due to the interference effect than if only one of the holes is open – just as we observe for the probability of electrons hitting the screen.

Particles and Waves

But how can this be? Electrons are particles and not waves! Waves are extended objects. Their amplitude and intensity can take on any value. An electron, on the other hand, always hits at a clearly defined point on the detector screen. Electrons always arrive in "identical lumps" like bullets, as Feynman puts it. But we can't explain the stripe pattern with bullets, while we can with waves. So it would be good if we could somehow combine the property of electrons to be "lumpy" with the interference ability of waves.

The following idea solves this problem. It is not the electron itself that is a wave. It is the probability of an electron hit that must behave like the intensity of a wave. So we have to assume that there is some kind of electron quantum wave whose intensity gives the hit probability. The more strongly the wave oscillates at some place, the more likely we are to find an electron there.

One more thing is important here, even though Feynman does not mention it specifically: it does not matter how many electrons are simultaneously on their way from the source to the detector screen. Even if the source only emits one electron every ten seconds, the result will remain the same. It's just that it would take a little longer for the pattern of stripes to build up from the slowly arriving electrons. This shows that electrons passing through one hole do not interact with electrons passing through the other, because there is never more than one electron on the way. Every electron interferes with itself! In 1989 a similar experiment was actually carried out by Akira Tonomura and his colleagues[1] (Fig. 5.7). There is also an impressive video on YouTube – just search for *One electron double slit experiment by Akira Tonomura.*

Feynman summarizes the result of the double-slit experiment, which is so important for quantum mechanics, in Sect. 1-5 of the third volume:

[1]See American Journal of Physics 57, 117 (1989).

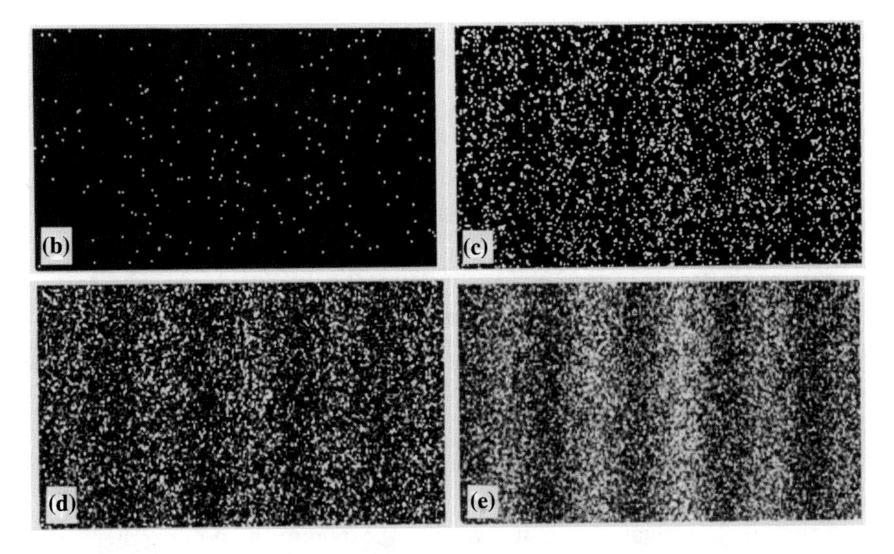

Fig. 5.7 Interference pattern of the double-slit experiment by Akira Tonomura with 200 (**b**), 6000 (**c**), 40,000 (**d**) and 140,000 (**e**) electrons. Here we see how the individual electrons gradually build up the typical interference pattern (© Akira Tonomura; CC BY-SA 3.0 Unported-Lizenz)

> The electrons arrive in lumps, like particles, and the probability of arrival of these lumps is distributed like the distribution of intensity of a wave. It is in this sense that an electron behaves sometimes *like a particle and sometimes like a wave.*

That is really all there is to say, Feynman concludes, admitting that there are still a lot of surprising subtleties hidden in this strange but very real behavior of nature. Here is an example of one such subtlety that leaves us humans rather at a loss.

Which Hole Does the Electron Pass Through?

Since electrons arrive in lumps at the detector screen, they should have passed through either hole one or hole two, by analogy with bullets. So Feynman makes the following proposition:

> Proposition A: Each electron either goes through hole 1 or it goes through hole 2.

If this were true, an electron would have two ways of reaching a certain point on the detector screen: it would have arrived either through hole one or through hole two. So the probabilities would have to add up – just as they would for bullets.

But they don't, as we have seen. With two open holes, the probability of an electron hitting the screen in a dark strip is smaller than with only one open hole. So we should undoubtedly conclude that proposition A is false, as Feynman writes: "It is not true that the electrons go either through hole 1 or hole 2."

This may seem strange, but it is logical! And it is interesting to ask what would happen if we were to use a tiny detector to register which hole the electron has passed through.

Proposition A must be correct now, because we actually *identify* the hole the electron passes through. So now the probabilities have to add up – and indeed they do! The pattern of stripes disappears and we get a rather fuzzy image of the two holes, as shown in the third bar of Fig. 5.6. There's no longer any interference!

If we remove the detector again so that we no longer know which hole each electron goes through, the stripes in the pattern will reappear and we may conclude once again that proposition A must be wrong.

The same goes for all methods that might be used to determine which hole the electron went through. They all lead to the same result. We can do whatever we want:

> It is impossible to design a piece of apparatus to determine which hole the electron passes through, without at the same time disturbing the electrons enough to destroy the interference pattern.

So what becomes of our proposition A? Is it true or not? Feynman says something like this in a slightly modified form: only if the passage of the electron through the holes leaves some trace in our macroscopic world that we can use at least in principle to determine through which hole the electron has passed, only then can we say that it goes through either hole 1 or hole 2. Otherwise we may not say that.

Well, let's say we don't use an electron detector and can't tell which hole the electron went through. Could the electron itself perhaps know where it has been, even if this information is inaccessible to us without a detector at the holes? Is there such a thing as hidden information about the location of the electron in nature? Perhaps the electron has some kind of inner workings that determine through which hole it goes and that we do not yet know about.

Feynman deals with this question as follows. This hidden information should not depend on whether we look at the holes or not with our detector. Whatever is inside the electron that determines which hole it goes through should not be dependent on what we do. Thus, even if we do not use the detector, there should be two groups of points in the pattern of the electrons on the screen: those that came through hole one and those that came through hole two. But then, as with bullets, the probabilities would have to add up and there would be no interference. So it looks as if this hidden information, often referred to as "hidden variables", does not exist – at least not in local form as a property of the individual electrons. So the electrons themselves do not know which hole they go through. Only when we take a look with a detector is it decided which of the two holes they go through, and then the probabilities do indeed add up. We will come back to this question in Sect. 6.2.

Nobody Understands Quantum Mechanics

But how does it work? How can all this be? "No one has found any machinery behind the law", says Feynman and continues: "No one can explain any more than we have just explained. [...] We do not know how to predict what would happen in a given circumstance, and we believe now that it is impossible – that the only thing that can be predicted is the probability of different events."

Feynman did everything he could to help people understand this central insight of modern science. A wonderful example is his lecture *Probability and Uncertainty – the Quantum Mechanical view of Nature*, which he gave in 1964 as part of the lecture series *The Character of Physical Law*, in the context of the Messenger Lectures at Cornell University. You can find the transcript in various places on the Internet. And even better, you can also find the original video recording of the presentation there and watch Feynman in action – just insert the title of the presentation in one of the common search engines.

Many parts of this talk correspond to the chapter on the double-slit experiment in the Feynman Lectures. But there are also some very nice formulations that can't be found in the Feynman Lectures. Feynman's "I think I can safely say that nobody understands quantum mechanics" is famous. He even warned his audience not to keep saying to themselves "But how can it be like that?", because that would lead down a blind alley from which nobody has yet escaped. "Just relax and enjoy it!" was his advice. This was typical of Feynman. In another talk he warned against being too rigorous too soon – that would lead to nothing: "Don't be so rigorous or you will not succeed".[2]

Another wonderful remark in the Messenger lecture is the following quip aimed at the despised philosophers: "A philosopher once said: *It is necessary for the very existence of science that the same conditions always produce the same results.* Well, they do not." That's exactly the way things are: in the double-slit experiment, under the same conditions the electrons randomly select either hole one or hole two – if we look at them – and then hit the detector screen somewhere with a certain probability. And yet this does not mean the end of physics as a natural science. According to Feynman, what is necessary for the existence of science is that minds should exist that do not insist that nature must satisfy certain preconceived conditions, like those of the above philosopher.

Feynman also repeats the above remark about philosophers in his well-known book *QED – The Strange Theory of Light and Matter*, which appeared in 1985 and is based on further talks given by Feynman. Many other ideas and formulations from earlier lectures can also be found there. We can see how over the years Feynman continued to build up his store of vivid explanations, helpful examples, and amusing asides. This finally enabled him to present even the difficult field of quantum electrodynamics to a broad audience in his book about QED, in such a way that we at least get the impression that we have understood the essentials.

How he achieves this is exemplary! All he would do is draw little arrows on a piece of paper – that's all! It sounds so very reassuring! These arrows are of course the probability amplitudes given by the quantum wave, which he also mentions in the appropriate places. Then he gradually develops the details we have already encountered: how arrows rotate with time like the hand on an imaginary stopwatch, what the squared arrow length has to do with probability, and why we have to add the arrows for different possibilities – for example, different paths – which leads to interference. "Draw amplitudes (i.e., arrows) for every way an event can happen and add them when you would have expected to add probabilities under ordinary circumstances", he writes, and later even makes it clear that this only applies if the different ways are indistinguishable in the particular experiment – if, for example, there is no way of knowing which hole the electron has gone through in the double-slit experiment.

---

[2]See H. D. Zeh: *Feynman's Interpretation of Quantum Theory.*

Simple examples such as the partial reflection of light on two glass surfaces, one behind the other, are all Feynman needs to illustrate this. Seemingly without any effort, he spins his red thread further and further through the thicket of details and arranges them into a single logical construction. Little by little, he lays the foundations for describing all QED processes with the help of his Feynman diagrams – in Chap. 3 of the present book, we have already got to know several of them. By the end of the approximately 170-page QED booklet, Feynman's didactic skill can easily make us think: "My God, it's not that difficult." There he is again: Feynman, the physics teacher who manages to cast a spell over almost everyone.

In his preface to the *Feynman Lectures*, Feynman stated self-critically that his presentation of quantum mechanics there had not been completely successful. Maybe he would have a chance to do it again someday. Then he would do it right. Well – about two decades later he finally succeeded in doing this with his presentation of QED.

5.4 Gravity and Quantum Theory

Feynman's lecturing activities at Caltech in the academic year 1962/63 were not limited to his famous Feynman Lectures for freshmen, which we got to know in the last section. Among other seminars, he gave a weekly lecture on gravitational theory for advanced graduate students and postdoctoral fellows who were already familiar with the methods of relativistic quantum field theory and Feynman diagrams. Due to the demanding topic, only about 15 people attended the lecture, which was held in a small room with just two rows of chairs.

Two of the postdocs, Fernando B. Morinigo and William G. Wagner, made transcripts of the lectures, which were later typed, duplicated, and offered as lecture notes at Caltech for many years. The script was highly sought after by many students and was widely distributed. Finally, it was revised again and published in book form with a very interesting foreword by John Preskill and Kip Thorne, as well as Brian Hatfield, under the title *Feynman Lectures on Gravitation*.

Feynman did not intend to give one of the usual introductory lectures on Einstein's general theory of relativity – that would have been far too boring for him. Once again, he preferred to follow his own path. He had already been thinking extensively about the theory of gravity for some time. The lectures would now help him to organise his thoughts and present them in an understandable way.

On the blackboard in Feynman's office was written his motto: "What I cannot create, I do not understand." This also applied to gravitational theory – so how would Feynman tackle this subject?

Why is Gravity So Weak?

The first successful attempt to establish a theory of gravity is more than 300 years old. In 1687 Isaac Newton published his famous law of gravity, which is shown in Infobox 5.1. According to this law, the gravitational force between two one-kilogram weights one meter apart is just 6.674×10^{-11} N. That is only about six hundred billionths of a newton – a tiny value that is very difficult to measure! No wonder we don't notice the gravitational attraction between pieces of furniture in our apartments, for example.

Infobox 5.1: Newton's law of gravity

Today we usually express Isaac Newton's law of gravity simply by the formula

$$F = G\frac{m_1 \cdot m_2}{r^2}.$$

Here F is the gravitational force between two masses m_1 and m_2 separated by a distance r from each other, and $G = 6.674 \times 10^{-11}$ N m²/kg² is the gravitational constant. For extended spherical masses like planets, we must use the distance r to the center. The gravity we feel on the Earth's surface is thus determined by the distance to the center of the earth (Fig. 5.8).

In words, Newton's law of gravity says the following. The gravitational force between two masses is attractive and acts along the connecting line between them. It decreases in inverse proportional to the square of the distance between the masses and increases in proportion to each of the two masses. Doubling one of the two masses thus doubles the gravitational force, while doubling the distance causes the force to drop to a quarter of its previous value. In order to calculate the exact value of the gravitational force, the gravitational constant G is also needed.

The weakness of gravity becomes even clearer when we compare the gravitational attraction between two protons with their electrical repulsion. The electrical force is stronger by about 36 orders of magnitude (i.e., by a factor of 10^{36})! Thus gravity is by far the weakest known interaction between elementary particles and usually plays no role at all there. One would have to increase the mass of a proton by 18 powers of ten – that is, by a factor of billions of billions – in order to make the gravitational force between protons

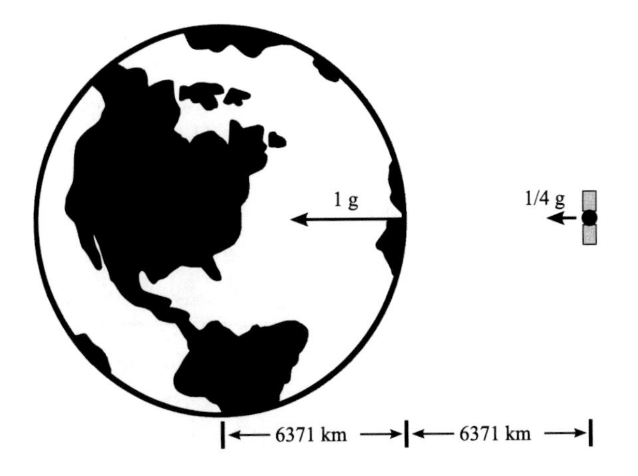

Fig. 5.8 A satellite two earth radii away from the Earth's center is only pulled down by a quarter of the gravitational force that would act on it at the Earth's surface (Author's graphics using http://freesvgs.blogspot.de/2016/04/happy-earth-day-free-svg-world.html)

as strong as the electrical force. This almost takes us into the range of the so-called Planck mass, which is about 1.3×10^{19} proton masses – see also Infobox 5.2.

Instead of asking why the gravity between elementary particles is so weak, we can also ask why the masses of these particles are so small that gravity between them plays practically no role. The two questions are equivalent, and nobody has yet found a satisfactory answer to either.

Feynman also addressed this question in his lecture. Instead of two protons, he used electrons, almost two thousand times lighter than protons, for the comparison, and found that gravity was 4×10^{42} times weaker than the electrical force. This number is so incredibly large that it is tempting to look for similarly large numbers that could be associated with it – an idea that goes back to Arthur Eddington, as Feynman observes. It is interesting to follow through Feynman's various speculations, although from today's perspective none of them has proved to be substantive. But who knows, maybe one day someone will find a way to explain the size of this number and thus the weakness of gravity. Fans of Douglas Adams' satirical novel *The Hitchhiker's Guide to the Galaxy*, for example, might notice the number 42 which appears in the exponent here. In the novel, the *Answer to the Ultimate Question of Life, The Universe, and Everything* is 42, as the huge supercomputer *Deep Thought* calculated after 7.5 million years. Too bad Douglas Adams later admitted: "It was a joke!"

But why is the effect of gravity so omnipresent in our environment when it is also so weak? The reason is that the force of gravity between particles is always attractive. So gravity can't be neutralized. In addition, like the electrical force, gravity has in principle an infinitely long range. Although it decreases quadratically with the distance, it does not decrease exponentially like the strong interaction, which is hardly noticeable even a few proton radii outside the atomic nucleus. Therefore, the gravitational effect of many particles can accumulate. But it takes the whole Earth to pull down one kilogram of iron with a moderately strong force of about ten newton. A single horseshoe-magnet, on the other hand, is enough to hold the piece of iron against the gravity of our whole planet!

The other three interactions in nature – the electromagnetic, strong, and weak interactions – are usually not as directly noticeable as gravity, although they are much stronger in the realm of particles. In the case of electrical forces, this is due to the fact that the influences of positive and negative charges tend to balance each other out when an object is viewed from the outside. The strong and weak interactions on the other hand have such a short range that they only play a role in the immediate vicinity of the associated particles. They are therefore essentially hidden deep inside atomic nuclei and hardly reach anything further afield. So the weakest of all interactions is the only one we feel directly on our own bodies and which was the first to be described mathematically, by Isaac Newton, over 300 years ago.

Einstein's General Theory of Relativity

Newton's law of gravity describes the effects of gravitation very accurately. Nevertheless it cannot be absolutely correct, because it assumes that gravity acts without any time delay between the masses. The current positions of the masses determine in which direction the gravitational attraction acts and how strong it is. This contradicts the special theory of relativity, according to which any physical action such as the action of forces – cannot propagate faster than the speed of light.

Einstein published his special theory of relativity in 1905, and since then it has been well corroborated in all areas of physics. So it was obvious that Newton's law of gravity would have to be changed in some way to take into account the delayed influence of gravity. Einstein realized this and had been working on this project since 1907.

He gradually developed several ingenious ideas to achieve his goal: a relativistic formulation of gravity. At first, however, it was unclear to him how he could formulate his ideas mathematically. He eventually found the appropriate mathematical language in what is known as differential geometry.

Fig. 5.9 Albert Einstein writes a fragment of his field equations on the blackboard. The photo was taken in Princeton in the winter semester 1930/31 (© picture alliance/akg images)

This had been developed in particular by the mathematicians Carl Friedrich Gauss and Bernhard Riemann in the early and middle 19th century. Even a genius like Einstein had first to familiarize himself somewhat laboriously with this new kind of mathematics. In doing so, he was supported by his friend Marcel Grossmann, who had meanwhile become a professor of mathematics in Zürich. In 1912, Einstein wrote to Arnold Sommerfeld, declaring that never before in his life had he toiled anywhere near as much, and that he had gained enormous respect for mathematics, whose more subtle parts he had considered until then, in his ignorance, as pure luxury.

With the help of differential geometry, Einstein could treat space and time as a single mathematical object – now called space-time – in which space and time are closely linked. He could then assign a curvature to this space-time, and this led to a relativistically correct description of gravity. In 1915 Einstein presented the field equations of his general theory of relativity as the final result of his years of drudgery (Fig. 5.9). With these equations he could describe how matter curves the surrounding space-time causing gravity, and thereby completed one of the greatest breakthroughs ever achieved by the human mind in theoretical physics.

Gravitons: The Quanta of Gravity

In his lecture, Feynman did not simply want to repeat Einstein's train of thought, which was already almost fifty years old by this time. Instead, he wondered whether it might not be interesting to begin where the majority of physicists now felt most at home: in particle physics and the associated quantum field theories, which Feynman himself had helped to shape with his Feynman diagrams. These methods had not been available to Einstein at the time. Feynman hoped that this new approach would yield further insights into the inner nature of gravity. Ideally, this would automatically result in a quantum theory of gravity, from which Einstein's general theory of relativity could be derived as a classical approximation. Moreover, the new approach could release gravitational theory from its isolation: physicists dealing with quantum theory and elementary particles usually ignored gravity and vice versa. There seemed to be no overlap between the two areas of physics. But if particle physicists could now find an access to gravity with their own methods, this artificial boundary would be removed and a first step would be taken toward including gravity in the quantum theoretical description of the world.

To illustrate his approach, Feynman – always the passionate storyteller – sketched the following scenario. Let us imagine that somewhere in the universe there is a highly developed alien civilization whose particle physicists know everything about quantum electrodynamics and the quantum physics of mesons and other particles. Gravity, on the other hand, is unknown to them – perhaps because there are no accumulations of masses such as stars or planets in their region of the universe whose gravity would be strong enough to be perceptible. But then something happens: a very sensitive new experiment shows that two large electrically neutral masses attract each other, albeit with only a very tiny force. This new force – i.e., gravity – had previously been missed by physicists due to its weakness.

What would the alien physicists – Feynman calls them Venusians – do to explain this new discovery? They would probably try to describe the new very weak attraction via a quantum field theory. This is exactly what they had succeeded in doing with electromagnetic force, for example – this interaction could be described by the exchange of photons. For gravity, corresponding gravitational quanta – or in short, gravitons – would be needed. These would then mediate the gravitation between massive objects, just as in QED photons are the mediators for the electromagnetic forces between charges.

Do We Actually Need a Quantum Theory of Gravity?

Of course, one can doubt whether one really needs a quantum theory to describe gravity. Do we really need gravitons, or isn't it enough to have the classical theory that Einstein developed? This question was asked several times in the first half of the twentieth century, and Feynman also addressed it in detail in his lecture. Are there any observable effects of gravity that only a quantum theory can explain?

In an atom, the gravitation between the electrons and the atomic nucleus plays no observable role in comparison to the electrical attraction, as Feynman showed. In the world of atoms, we have no chance of ever seeing the effects of gravity. The masses of electrons and atomic nuclei are simply far too small for significant gravitational forces to develop between them. Other quantum effects of gravity – for example, the effects of individual gravitons – are also very tiny and hardly measurable.

Maybe we should not try to quantize gravity, Feynman asked. Is it possible that gravity is not quantized and all the rest of the world is? Is the classical description Albert Einstein found with his general theory of relativity sufficient?

The following theoretical argument speaks against this idea. In quantum theory, states can be constructed for a particle – say an electron – in which it is simultaneously at two locations, for example at the two holes in the double-slit experiment. Then what about its gravitational field? If there were only one classical gravitational field, the gravity of the electron could at least in principle indicate at which hole it is located. The quantum mechanical two-position state of the electron would immediately collapse and the electron would have to choose a location – either at hole one or at hole two. But then there would never be interference on the detector screen behind the double slit, because for this the two possibilities – passage through hole one or two – must be indistinguishable.

In order for such two-position quantum states of the electron to exist, the associated gravitational field of the electron must also be in a suitable quantum mechanical state – i.e., in a superposition state of two gravitational fields belonging to the two locations of the electron. In short, if the location of a particle becomes undetermined and fuzzy due to quantum mechanics, this must also apply to the gravitational field of this particle. And since Einstein's theory of general relativity describes gravity as the result of the curvature of space-time, it follows that space-time itself must somehow become fuzzy and quantum-mechanically blurred – a fascinating thought!

This argument is convincing enough and generally accepted – as it was also by Feynman. Nevertheless, Feynman urges caution and advises us to keep an open mind – after all, the gravitational field of an electron is far too small ever to be measurable.

Are there any phenomena in the universe that could not exist without a quantum theory of gravity? This is not the case for the usual phenomena of our world. Where quantum mechanics is important – i.e., in considerations of atoms and particles – gravity is hardly perceptible. And where gravity is important – i.e., in considerations of the stars and planets – we do not need quantum mechanics. Only when large masses and very small space and time intervals are both involved do the two theories become relevant at the same time.

The so-called Planck units, which are described in Infobox 5.2, provide a good indication of when this will be the case. Such extreme dimensions can only be found in the Big Bang or inside black holes. So this is exactly where we will need quantum gravity. By the way, on the Internet there is a very nice cartoon video about this, made by the theoretical physicist Benjamin Bahr – just search for *Quantum Gravity in under five minutes.*

Infobox 5.2: Planck units

In quantum theory, the Planck constant h or $\hbar = h/(2\pi)$ is a new physical constant that relates the particle properties energy E and momentum p to the wave properties frequency f and wavelength λ via the formulae $E = h \cdot f$ and $\lambda = h/p$.

In the special theory of relativity, there is also a universal constant of nature: the speed of light c. It connects spatial and temporal separations, but also, e.g., energies and masses ($E = mc^2$).

In relativistic quantum field theories such as QED, which respect the laws of both quantum mechanics and special relativity theory, the two physical constants $\hbar$ and c must therefore both occur.

If gravity is included, the gravitational constant G must also occur, as a third physical constant, in fact a measure of the strength of the gravitational force. From these three physical constants the following quantities can be obtained:

$$\text{Planck mass: } m_P = \sqrt{\frac{\hbar c}{G}} = 1.3 \times 10^{19} \text{ u}$$

$$\text{Planck length: } l_P = c\,t_P = 1.6 \times 10^{-35}\ \text{m}$$

$$\text{Planck time: } t_P = \frac{\hbar}{E_P} = 5.4 \times 10^{-44}\ \text{s}$$

$$\text{Planck energy: } E_P = m_P c^2 = 1.2 \times 10^{19}\ \text{GeV}$$

Here the unified atomic mass unit "u" represents approximately the mass of a proton or a neutron.

These Planck quantities constitute natural units for masses, energies, times, and lengths that are independent of man-made units such as meters or kilograms. At the same time, the Planck quantities show on which scales general relativity theory and quantum theory become equally important, so that a theory of quantum gravity would be needed. In the quantum world of particles, this is only the case at extremely short distances and time intervals and very large particle masses and particle energies, situations which are all far beyond our experimental abilities. For example, the Planck length is about 20 orders of magnitude smaller than the diameter of a proton, while the Planck mass is about 19 powers of ten greater than the proton mass.

From the Spin of Gravitons to Einstein's Field Equations

Now let's return to Feynman's initial question: what would the alien physicists do with all their knowledge of quantum field theory to describe the newly discovered weak attraction between masses? Which properties would gravitons have to have as the mediating particles of gravity in order to explain the experimental results?

Accurate measurements by the alien physicists first show that gravity decreases quadratically with increasing distance. This is the same as for the electric force and implies that gravitons must be massless, analogously to photons. If they had a mass, gravity would decrease much faster – Yukawa had used precisely this idea to explain the short range of the strong nuclear forces by the exchange of massive mesons (quarks and gluons were still unknown at that time).

Next, we must identify the spin of the gravitons, i.e., their quantum mechanical intrinsic angular momentum. According to the rules of quantum mechanics, the spin may take only half-integer or integer values. Feynman explores all possibilities one after the other.

Half-integer spin for the graviton – for example spin 1/2 as for electrons or neutrinos – would disallow the possibility of a static gravitational force between masses. Quite generally, therefore, fermions cannot be the exchange particles of an interaction. The reason for this is quite subtle and Feynman explains it using his Feynman diagrams – we prefer to skip the details here. As for all other particles mediating an interaction, the spin of the graviton must therefore be an integer.

Odd integer spins – i.e., 1, 3, 5 etc. – would lead to repulsive gravitational forces between identical particles. For example, photons with their spin 1 cause a repulsive electrical force between charges of the same sign. Such spins are also out of the question, because the gravity between particles is always attractive. So only even integer spins 0, 2, 4, etc., remain.

In the next step, Feynman also rejects spin 0 for gravitons. It would imply that gravity decreases as the mass of an object increases. But this should be the other way around. In his preface to Feynman's lecture, Brian Hatfield makes another argument, which he proves mathematically: if gravitons had spin zero, they would not interact with photons. So light would not then be subject to the influence of gravity, which would contradict observation.

In the case of gravitons with spin 2, the last issue finally works: as the mass increases, so does gravity. But does the rest work?

Feynman investigated this question in detail and checked step by step whether the resulting quantum theory of gravity leads to Einstein's general theory of relativity, just as photons lead to the Maxwell equations of the electromagnetic field. That is indeed the case, as Feynman can demonstrate. The details are too complex to be presented in this book – but on the other hand, there is no witchcraft here, as Feynman points out: we only need to follow a succession of logical steps guessed at by analogy with other theories such as QED.

How much more difficult it had been for Einstein to reach the same goal without these advanced tools. Feynman admits that he had no idea how Einstein had ever guessed at the final result. Feynman had already had to overcome enough difficulties in his own derivation starting from quantum field theory, but he had the feeling that Einstein had achieved the same under much more difficult conditions: swimming underwater, blindfolded, and with his hands tied behind his back. Max Planck warned Einstein at the time, when he was becoming increasingly entangled in his equations: "As an older friend I must advise you against it, for in the first place you will not succeed, and even if you succeed, no one will believe you." Well, Einstein persevered and finally succeeded, and a few years later the whole world believed him.

Feynman was convinced that the alien particle physicists would also very likely have used their methods from quantum field theory and discovered that gravity could be described by massless spin-2 gravitons. In cases where quantum effects do not play any role, they could have derived the equations of general relativity, just as Feynman did. Thus Feynman had actually achieved his goal and developed the theory of gravity in his own way.

As John Preskill and Kip Thorne explain in the preface to Feynman's gravity lectures, Feynman was not the first to come up with this idea. However, it is difficult to say to what extent he was aware of the work done by his colleagues on this subject. There is much to suggest that he worked out his ideas by himself, as he so often had in the past.

Theorists would address this topic again and again later on. In 1964, Steven Weinberg, who later won the Nobel Prize, found a particularly beautiful approach. Weinberg was 15 years younger than Feynman and thus already belonged to the next generation of physicists. In his derivation he started from the known fact that the gravitational field also has an energy and thus itself causes gravity again. A gravitational field thus acts back on itself. In the language of quantum theory, this means that gravitons – unlike photons – interact and influence each other directly. Weinberg deduced from this that the corresponding theory of spin-2 gravitons obeys the rules of special relativity only if the gravitons interact with all other particles and the strength of the interaction does not depend on the particle type, whence gravity affects any particle in the same way. From this, Einstein's field equations can ultimately be derived.

The Total Energy of the Universe

Once Feynman had derived Einstein's general theory of relativity in his very own way, he also dealt in detail with the consequences of this theory. And he noticed something remarkable: the total energy of the universe could actually be zero!

At least this follows if we consider the density and expansion rate of the universe, which were only approximately known at that time. Negative gravitational energy and the positive energy contained in the mass of matter then have the same order of magnitude. So the negative gravitational energy of the universe could just balance the positive energy contained in its mass. The additional gravity of a new particle could provide just as much energy as is needed to create the particle. It is exciting to think that it might cost nothing to create a new particle, says Feynman, and continues: "Why this should be so is one of the great mysteries and therefore one of the important questions of physics."

In Einstein's theory of gravity, however, the universe only has total energy zero if the matter in it has a certain average density, which is called the *critical density*. Modern measurements show that the universe does appear to have exactly this critical density, which corresponds to a mass of just under six hydrogen atoms per cubic meter – although of course it does not necessarily have to consist of hydrogen atoms alone.

In Feynman's day, measurements were still a long way from this level of accuracy. However, the inaccuracy in the values for the density of the universe at that time did at least make it possible that the universe could have exactly the critical density. Feynman noted that the critical density was just about the best density to use in cosmological problems for the time being. It has a great many pleasing properties – e.g., the universe does not on average exhibit any space curvature at critical density. It seemed exciting to speculate that the critical density might indeed be the actual density. On the other hand, Feynman reminds us that we must not fool ourselves into thinking that a beautiful result is more reliable simply because of its "beauty".

Feynman would certainly have been thrilled that the methods of modern astronomy – especially by measuring the small irregularities in the cosmic background radiation with the space observatories WMAP and Planck – have made it possible to determine the mean density of the universe quite accurately, and that the critical density is the value that actually comes out!

So it looks as if our universe may actually have zero total energy. And if we need practically no energy to create our universe, that would make it something like the ultimate free lunch – a fascinating thought!

Wanted: A Quantum Theory of Gravity

Of course, in his lecture Feynman was not only concerned with the classical theory of gravity, which Albert Einstein had already found. The spin-2 gravitons were quantum mechanical objects and should therefore provide a basis for a quantum theory of gravity – or so Feynman hoped.

Feynman's roadmap for his further studies of quantum gravity was quite simple: use massless spin-2 gravitons and then work out the rules according to which the corresponding Feynman diagrams could be translated into formulas. These rules would be used to calculate first simple and then more complex diagrams, just to see what happens.

In quantum electrodynamics (QED), Feynman's approach had been similar and had ultimately been successful. However, there had already been certain problems there. As soon as the diagrams become complex enough to contain closed loops, the corresponding formulas result in an infinite contribution. The infinity has its origin in the fact that the particle energies in the loops can become arbitrarily large. If all these energies are taken into account in the integrals, the result is infinitely large.

In Chap. 3 we have seen how the problem can be solved in QED: we hide the infinities in the terms for the physical charge and mass of the electron. We then require that they be compensated by a corresponding counter-behaviour in the bare electron charge and mass, in such a way that the measured values for the physical charge and mass are obtained.

In QED this trick, known as renormalization, works perfectly. It ensures that even complex diagrams with many loops and vertices nevertheless make sense and can be used to calculate the physical processes to high accuracy. So QED can be *renormalized.*

Does the same work with gravitons? Feynman was unable to answer this question, but he doubted this, and surmised in his lecture that his theory of quantum gravity would not be renormalizable. However, he did not know whether this was a significant objection to any theory – even in QED, Feynman considered renormalization as merely an elegant method for sweeping the difficulties with the divergences under the rug.

In fact, it has turned out that the quantum theory of gravity cannot be renormalized if it is constructed with the established standard methods, as Feynman had tried. The infinities arising in the more complex diagrams can no longer be hidden in just a few measured physical parameters, such as particle masses. More and more parameters are needed as the diagrams become more complex. The more accurately we want to calculate a gravitational quantum process, the more diagrams and thus the more measured parameters we have to put in. To achieve perfect accuracy, i.e., by including an infinite number of diagrams, an infinite number of measured parameters must be used, and that makes the whole theory absurd.

In the past, therefore, the lack of renormalizability was often seen as a sign that a quantum theory of gravity based on gravitons made no sense, because it would not be well defined. Today, this conclusion is viewed with more caution.

If we limit ourselves to relatively simple diagrams as Feynman did in his lecture, then we can also achieve meaningful results in quantum gravity. The infinities are therefore viewed today rather as a sign that quantum gravity based on gravitons only makes sense up to certain particle energies. Feynman was probably right, therefore, not to consider renormalizability as being so important, at least for this range of energies.

On the other hand, a theory of quantum gravity is particularly relevant at the very highest energies, such as in the center of black holes or in the Big Bang, so the question of renormalization has to be raised again here. The search for a renormalizable theory of quantum gravitation is equivalent to the question of what really happens at these extreme particle energies – or equivalently, very short distances. What happens in nature when particle energies approach the Planck energy and wavelengths and distances approach the Planck length?

This is exactly the question many physicists are working on today. There are certainly some interesting approaches around, including for example string theory and loop quantum gravity. These theories no longer work with point-like electrons, photons, and gravitons, which can be represented as lines in Feynman diagrams. String theory, for example, is based on tiny vibrating strings with lengths in the range of the Planck length. In today's experiments we cannot resolve the tiny spatial extensions of these strings, so they appear to us like point-like particles whose properties are generated by the vibrations of the string.

Interestingly, in string theory, there inevitably arises a string oscillation that looks from a distance like a massless spin-2 particle, i.e. like a graviton. Thus, string theory automatically contains a quantum theory of gravity. At the same time, many of the infinities that occur with point-like gravitons are avoided due to the tiny spatial extension of the strings.

That sounds quite promising at first. But string theory also has its problems, and other approaches are also being intensively investigated. One thing is clear now: we will only get a comprehensive and consistent picture of nature when we can include gravity in the quantum description of the world. Richard Feynman followed this path at a time when many of his colleagues still considered it to be a rather exotic project. That was just right for Feynman – he liked to move off the beaten track. In the meantime, the search for a comprehensive quantum theory of gravity has become one of the main topics in theoretical physics, and many physicists are looking for ways to solve the critical problems that arise in this quest.

6

Quarks, Computers, and the Challenger Disaster

In the 1960s, there were increasing signs that protons, neutrons, and ultimately all strongly interacting particles were composed of smaller constituents. But there was no experimental proof for these constituents, which Murray Gell-Mann called quarks, so many physicists had doubts about their existence. In this context, Feynman made his last major contribution to physical research. In 1968, he showed how the high-energy collisions of electrons on protons could be interpreted as if the electrons had hit smaller particles inside the protons. This was the breakthrough, and soon no one doubted the existence of quarks any more.

This discovery was followed by dramatic progress that culminated in the so-called Standard Model of particle physics. Feynman followed the proceedings with interest, but no longer made any decisive contributions. Younger physicists such as Steven Weinberg, Abdus Salam, Sheldon Glashow, and many others had since taken the helm. Moreover, Feynman fell ill with cancer for the first time in 1978, and this increasingly affected the last decade of his life.

However, his illness did not prevent him from continuing to lead an active life. He was very interested in computers and their possibilities, lectured on them, consulted a start-up computer company, and investigated whether it was possible to learn something about physics from computer science. Is the universe perhaps itself something like a huge computer?

In 1986, he got involved in something he had always avoided before: he became a member of the official commission to investigate the Challenger Space Shuttle explosion. There was an unforgettable moment when he

J. Resag, *Feynman and His Physics*, Springer Biographies,
https://doi.org/10.1007/978-3-319-96836-0_6

demonstrated with a glass of ice water in front of running cameras where the cause of the disaster could be found.

Richard Feynman died in Los Angeles on February 15, 1988 – he finally lost his long battle against cancer.

6.1 Symmetries and Quarks

Since his last great scientific success in particle physics – the description of parity violation in 1957 – Feynman had largely withdrawn from this field. Perhaps it had simply become too crowded for him in this area of physics, which has by now become one of the most important. Ever more powerful accelerators were producing more and more data, and whole legions of physicists were trying to discover some kind of order in this tangle of new particles and their properties. Feynman rather pursued his own interests. He wanted to solve interesting problems – preferably alone – and was not interested in reading the countless papers published by his colleagues. Feynman's work on quantum gravity is a good example of his way of avoiding the beaten track.

In addition, Feynman's life in the early 1960s was full of things that demanded his attention. Finally, after his marriage to Gweneth in 1960, his private life had become more serene. His son Carl was born in 1962, and in 1968 they adopted a little girl called Michelle.

In addition to his family activities, he was greatly occupied by the *Feynman Lectures*, which required his entire commitment between 1961 and 1963 (see Sect. 5.3). In parallel with this he gave his lectures on gravity, and at the end of the two lecture series he was thoroughly exhausted. Many other talks and lectures also came along, such as the six messenger lectures in 1964. And of course he was awarded the Nobel Prize in Physics in 1965.

Symmetries in Physics

So Feynman was busy with a lot of things in the early 1960s, and it was not until a few years later that he became increasingly involved in physical science again. But of course, research in particle physics also went ahead without him. Feynman's sparring colleague and competitor at Caltech, Murray Gell-Mann, played a major role in this. He was at the height of his scientific career and tackled the problem of bringing some sort of mathematical structure to the zoo of particles.

The most important tool he introduced was the branch of mathematics known as group theory. It can be used to describe symmetries in physics in an elegant way, and thus identify similarities among the particles. An example of classical symmetry is rotational symmetry: if we rotate a three-dimensional sphere around its center, its appearance does not change – it is spherically symmetrical. The rotations in three-dimensional space thus form a symmetry group for the sphere – there is more about the mathematical concept of a group in Infobox 6.1.

Infobox 6.1: The mathematical concept of a group

The rotations in three-dimensional space form a group because they have the following mathematical structure:

- two rotations executed one after the other result in another rotation,
- there is a counter-rotation (called an inverse element) for each rotation, which undoes its effect,
- there is a rotation (called the neutral element) that does not change anything, namely the rotation through zero degrees (about any axis), and
- with three rotations in a row, the first two or the second two rotations can be combined to a new rotation before they are combined with the third rotation (associative property).

Any mathematical entity that satisfies these laws is called a group – the entities do not necessarily have to be rotations. For example, the integers under addition also have the properties of a group: $1+2$ is also an integer, $1+(-1)=0$, $1+0=1$ and $(1+2)+3=3+3=1+(2+3)=1+5$.

By the way, one thing we don't necessarily require of a group is commutativity. It can make a difference which rotation we do first, and in many cases it does! While the group of displacements (translations) in space or the addition of integers allows us to swap the order in which the group entities are combined, with spatial rotations around several axes the sequence is quite important (would you like to try it?).

What happens if we don't rotate a sphere, but an atom?

Let us assume that the standing electron wave in the atom oscillates in such a way that there is an antinode at the top and bottom. In Fig. 3.13 of Sect. 3.3 we got to know this waveform as the $2p_x$ orbital. This standing wave is dumb-bell-shaped with two lobes pointing in opposite directions from each other along the vertical axis, i.e., the electron is likely to be found above or below the atomic nucleus. If we now turn or tilt such an atom, we also tilt the standing electron wave. For example, by tilting the $2p_x$ orbital by 90°, we can arrange for the lobes to lie to the right and left, whereupon the $2p_x$ orbital becomes a $2p_z$ orbital.

What does not change when tilting the orbital in this way is the energy of the electron. For the energy, it doesn't matter how the lobes are oriented in space, as long as the standing electron wave keeps the same shape.

Rotating or tilting thus changes the spatial orientation of the quantum wave of the electron, but not its energy. So all electron quantum waves of an atom that can be converted into each other by a rotation of the atom – in our example, the $2p_x$, $2p_y$, and $2p_z$ orbitals – have the same energy.

Gell-Mann's Eightfold Way – A Periodic Table of Particles

This brings us to the crucial point of our symmetry considerations: if a symmetry transformation – for example, a rotation – converts different quantum states into each other, these quantum states must have the same energy.

Can we possibly use this idea to look for corresponding relations in the zoo of particles? For example, are there particles with the same or similar energy which, according to Einstein's formula $E = m \cdot c^2$, correspond to the same or similar particle mass? In the following we will limit ourselves to the strongly interacting particles – baryons and mesons – because they represent the majority of all species in the particle zoo.

In Sect. 4.3, we saw that this is indeed the case: the three pions π^+, π^-, and π^0 are about equal in weight, with masses of just under 140 MeV, as are the three rho mesons ρ^+, ρ^-, and ρ^0, which have masses of about 770 MeV. Is there perhaps some kind of abstract "rotation" that can, for example, transform the three pions into one another and perhaps turn a π^+ into a π^- without significantly changing its energy or mass? A spatial rotation is obviously out of the question: a positively charged π^+ remains a π^+, even if we turn ourselves upside down.

Nevertheless, there is something to the idea – we just have to adapt it somewhat. Suppose the pions consist of some kind of constituents, with each of the three pions having a different composition. For example, if we now replace certain constituents with others in the positive pion π^+, this pion could become a negative π^- or a neutral π^0. The corresponding quantum states would be transformed into each other, as with our rotated atom.

Let's also imagine that some force brings constituents of equal mass together to form the pions. If this force doesn't care which constituents it is dealing with, we could permute the constituents without any change in the binding energy and thus the mass – all three pions would have the same mass, as we know they do in reality.

In the 1960s, in addition to the three pions and the two nucleons (proton and neutron), many other mesons and baryons were already known. Gell-Mann was looking to see whether he could also find particle groups with the same or at least similar masses among them. Then he thought about how many different constituents he would need to explain these particle groups. It looked as if three different constituents would suffice. These constituents are also said to come in three different *flavors*, although of course this has nothing to do with a real flavor like salty or sweet. He called these three flavors u (up), d (down), and s (strange). The last term in particular was no coincidence, because Gell-Mann had recognized that the constituent with the flavor "strange" had something to do with the strangeness he had introduced to explain the strangely long lifetime of the kaons, for example.

For the mesons, Gell-Mann had to combine two constituents – or more precisely, a constituent and an anti-constituent, i.e., a particle and an antiparticle. From the three flavors u, d, and s and the corresponding antiflavors $\bar{u}$, $\bar{d}$, and $\bar{s}$, he was able to generate a total of $3 \cdot 3 = 9$ flavor–anti-flavor combinations, these being divided into a group of eight interconnected states (also called an octet) and a separate state (also called a singlet).

The above considerations are somewhat simplified. The origin of these two groups of particles is connected with the fact that we do not only exchange constituents in order to move from one meson to another. In detail, the whole matter is more complicated, because we need something like a generalization of rotations that exchanges flavors, i.e., a kind of rotation with the three flavors taking the place of the three directions in space. Mathematically, the corresponding group is called *SU(3)*, where the 3 in brackets corresponds to the three flavors.

With this method Gell-Mann succeeded in assigning all mesons known at that time to either a set of eight states (octet) or a single state (singlet), in such a way that the mesons in their appropriate octet all had at least roughly similar masses. The octet of the lightest mesons is shown in Fig. 6.1 on the left.

In this way, Gell-Mann was able to establish a first structure in the confusing particle zoo. And his method was not only successful with the mesons. Instead of combining flavor and antiflavor to form mesons, he also made combinations of three flavors and was able to divide the baryons into particle multiplets with rather similar masses. The lightest baryons, for example, can also be assigned to a set of eight flavor states (see Fig. 6.1 right).

Clearly, the number *eight* plays a special role here. Gell-Mann repeatedly succeeded in introducing memorable terms into physics. He called his method the *Eightfold Way* in 1961. In doing so, he was alluding to Buddhism, in which the Noble Eightfold Path is one of the principal

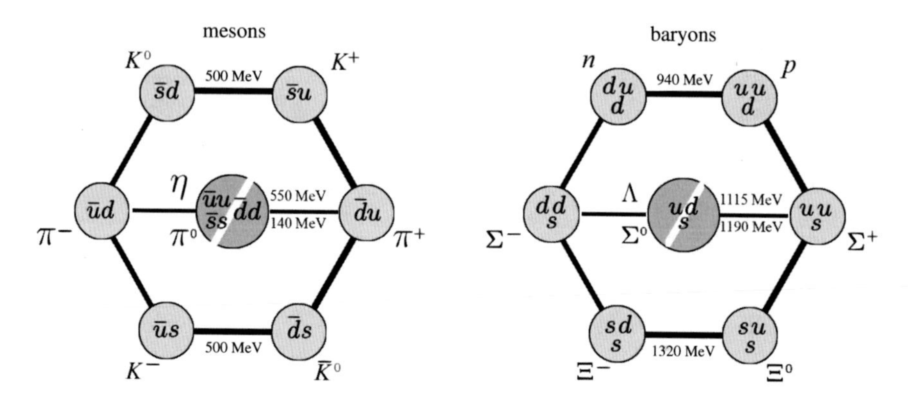

Fig. 6.1 The octets of the lightest mesons and baryons

teachings. Gell-Mann has an enormous general knowledge and loves such linguistic allusions. Incidentally, the Israeli physicist Juval Ne'eman formulated a similar classification of mesons and baryons at almost the same time, independently of Gell-Mann. The idea was obviously in the air!

However, there do not always have to be eight states which form a common group with similar mass. The three flavors can also be combined to form a group of ten (a so-called decuplet), which are permuted through flavor rotation. This decuplet also contains three states each comprising three identical flavors, i.e., *uuu*, *ddd*, and *sss*.

With his method, Gell-Mann was able to place every meson and baryon known at that time somewhere in a set of related particles. Almost a hundred years earlier, something similar had been achieved with the chemical elements. In 1869 the Russian chemist Dmitri Mendeleev and independently the German chemist Lothar Meyer had created the Periodic Table of the Elements, in which each element has its unique place. There were also gaps in this system. Some chemical elements predicted by the periodic table were missing. At that time they were still unknown, for example, the elements now called gallium, scandium, and germanium. All the missing elements were later found one after the other, and so the periodic table passed its first test.

There was just such a gap in Gell-Mann's arrangement of the particles: the particle with flavor combination *sss* was missing. This particle, known as Ω^- (the omega minus baryon), was actually found at Brookhaven National Laboratory in 1964! Gell-Mann's classification scheme had thus fulfilled an important characteristic of a good theory: it not only classified already known phenomena, but it also made predictions which were later confirmed by experiments.

Quarks: Constituents of Baryons and Mesons

So Gell-Mann's system was very successful and provided a clear indication that mesons and baryons were indeed composite particles. When it came to naming the hypothetical constituents in 1964, Gell-Mann's linguistic talent came to the fore once again. At first, he had a word in his mind that was supposed to sound like *kwork*. Then he came across the following lines in the novel *Finnegans Wake* by the Irish author *James Joyce*:

> Three quarks for Muster Mark!
> Sure he hasn't got much of a bark
> And sure any he has it's all beside the mark.

Three *quarks* – that was exactly what Gell-Mann needed for his Eightfold Way! As you can see, Finnegans Wake does not belong to the world's most easily digestible literature. It contains many linguistic experiments, often with words from other languages. It is quite possible that Joyce had read the word *quark* in a German dictionary (in Germany, "Quark" is similar to curd cheese). It is also said that Joyce picked up the word in the German city of Freiburg when market women offered him their dairy products – if that's true, it would be a funny coincidence! And since *quark* seems to rhyme with *Mark* and *bark* in the above lines, Gell-Mann revised the pronunciation of *quark* accordingly – the correct pronunciation of words was always an important concern to him, unlike Feynman, who hardly cared.

So what properties do quarks have to have if they are to be the constituents of mesons and baryons? What mass, charge, and spin do they have to possess?

The question of mass is not so easy to answer, because the term *mass* is actually only defined for particles which can move freely. However, no one has ever discovered free quarks. They seem to be locked inside the baryons and mesons – a phenomenon called *confinement* (more on this later in this section). The mass of a baryon or meson is not simply the sum of the quark masses, but is mainly the result of the large binding energies between the quarks.

Nevertheless, at least one thing is clearly visible in the meson and baryon masses, as shown in Fig. 6.1: the more strange quarks in a state within an octet or decuplet, the more massive it is. So it looks as if up and down quarks are about the same weight, while the strange quarks have more mass.

What about the electric charges of the quarks? What electrical charge has to be assigned to each quark flavor to explain the charges of the mesons and baryons? This is relatively easy to see:

The charge +2/3 must be assigned to the up quark, and $-1/3$ to the down and strange quarks, with the antiquarks carrying the opposite charges. The proton (uud) and neutron (udd) thus carry the charges $2/3+2/3-1/3=1$ and $2/3-1/3-1/3=0$, respectively, while for the positive pion ($u\bar{d}$) the charge is $2/3+1/3=1$. So when it comes to charges, everything works fine.

Are Quarks Real?

The fractional charge values of the quarks were very unusual – nobody had ever discovered a particle with an exotic charge like that. Such particles would have been easily visible in experiments. Instead, only particles with integer elementary charges were ever seen – a circumstance that is still not fully understood even today. So there was not the slightest indication that particles with fractional charges could even exist. Accordingly, the vast majority of physicists were suspicious when fractional charges were required for the quarks. Could such weird particles exist? Why was there no direct indication of such bizarre charge values? Was Gell-Mann's mathematically motivated model merely an abstract group-theoretical gimmick without a real physical foundation?

Clearly, there are strange things going on when we ask about the properties of quarks. What about the spin? Since mesons have integer spin and baryons have half integer spin, the quarks must have spin 1/2, like electrons. In the mesons, the spins of the quark and antiquark can be combined to a total spin of 0 or 1, while in the baryons the three spins can be combined to a total spin of 1/2 or 3/2.

But now there is a problem. In the Ω^- baryon, which consists of three strange quarks, all three quark spins are aligned parallel to each other. The three quarks with the flavor combination *sss* and the spins $\uparrow\uparrow\uparrow$ therefore appear to occupy identical quantum states, which is forbidden for fermions such as quarks, according to the Pauli principle.

Could it be that the Pauli principle is violated in quarks? This was out of the question for most physicists, because the Pauli principle was a fundamental quantum-mechanical law which followed inevitably from the combination of quantum mechanics and relativity theory. Before abandoning the Pauli principle, they would rather do without quarks.

So there were good reasons to doubt the real existence of quarks. The great majority of physicists preferred to stick to the previous idea that baryons and mesons should be spatially extended, but nevertheless not composite objects. There were even theories circulating, such as the so-called *bootstrap model*, according to which there should be no fundamental objects among these particles at all. Instead, in a self-consistent system, all baryons and mesons should be mutually equivalent building blocks, so that there would no longer be any fundamental difference between elementary and composite particles – one also spoke of *nuclear democracy*. The whole system of the various mesons and baryons should pull itself up by its bootstraps.

The young Russian–American physicist George Zweig had direct experience of how strongly physicists were opposed to the idea of quarks. Zweig had been to Caltech in the early 1960s and asked Murray Gell-Mann if he would be willing to supervise his doctoral thesis. Unlike Feynman, Gell-Mann usually supervised many doctoral students, but this time he had to refuse because he wanted to take a sabbatical year. However, he promised to ask Dick, as he usually called Feynman. At the time, Zweig was attending Feynman's lectures on gravitation, so there was hope that Feynman would at least remember Zweig's face. Hence, Zweig asked Feynman, and Feynman said that if Murray thought Zweig was okay, he must be okay. This allowed Zweig to enjoy many interesting exchanges in which he could discuss his ideas with Feynman.

One of these ideas was the following. Zweig had noticed that the so-called φ meson only rarely decays into three pions. It decays significantly more often into two kaons, i.e., mesons with strangeness, while pions have no strangeness. Zweig suggested that this could be because the φ meson was made up of different components to the pions, while the components of the φ meson could be found in the kaons. Today we know this to be true: the φ meson consists of a strange quark and a strange antiquark ($s\bar{s}$). During decay, this quark pair can easily divide into two kaons, so this decay is preferred.

When Zweig asked Feynman for his opinion, he reacted with reserve. He doubted the experimental data on the decay of the φ meson because it did not fit into a certain theoretical scheme that was considered indispensable at the time – we should remember that, since his experience with the inaccurate data on the beta decay of the neutron, Feynman had decided never to trust any experiments or expert opinions without careful consideration. But this time Feynman was wrong: The data was reliable, while the understanding of the strong interaction on which Feynman's objection was based at the time was wrong.

Zweig insisted on his idea and called the constituents *aces*. Since these aces seemed to have dynamic properties such as mass, binding energy, and spin, Zweig was convinced that they must be real particles inside the mesons and baryons – here Gell-Mann was much more cautious about his quarks and did not yet dare to venture so far. When Zweig wanted to publish his idea at the European research center CERN near Geneva in 1964, he met with resistance from the head of the theory group there, who prevented publication of the manuscript. Quarks or aces were regarded by many as useful mathematical tools for classifying mesons and baryons, but were not taken too seriously as real physical objects. So there it is again: the recurrent discussion about what should be considered *real* in physics.

Feynman and the Search for Quarks

So there was a lot going on in particle physics back then. Initially, Feynman did not participate in the discussions about the existence of quarks. Moreover, the group theory on which Gell-Mann's model was based was not one of the tools Feynman usually used – unlike Feynman, Gell-Mann loved abstract mathematical tools. Feynman preferred a hands-on approach to physics, based on visual ideas and experimental facts. So he wondered whether one could learn more about the inner life of baryons and mesons from the available experimental collision data. If there really were smaller constituents inside these particles, it might be possible to establish this by the way the particles behaved during collisions, bouncing off each other or producing further particles from the available energy.

However, the available collision data did not provide any real indication of a substructure for these particles. For example, when protons bounced off each other elastically like rubber balls, without generating further particles, they behaved like objects with a diffusely smeared charge distribution that did not have any substructure. If, on the other hand, the collision energy was sufficient to generate further particles, the situation quickly became too confusing to draw clean conclusions.

In order to illuminate the inside of a proton as cleanly as possible, it is best to bombard it with a particle that is itself point-like, with no substructure of its own. The ideal particle for this is the electron. So what is needed is a kind of super-electron microscope for protons.

However, normal electron microscopes are a very long way from attaining the necessary resolution. The electrons used therein have much too low an energy, so their quantum mechanical wavelength is far too long to be able to see the inside of a proton. This requires electrons with very high energy that only a particle accelerator can generate. Exactly such an accelerator for electrons had been built in 1962, close to Stanford University near San José in California – it was called the *Stanford Linear Accelerator* or *SLAC* for short

Fig. 6.2 Aerial photo of the Stanford Linear Accelerator (SLAC) around 1968. On the left is the approximately 3 km long linear beamline used to accelerate the electrons. The buildings for the particle detectors are in the foreground at the bottom left Source: U.S. Department of Energy

(Fig. 6.2). In 1966, it went into operation and accelerated electrons with a 3 km long beamline to a final energy of around 20 GeV – about twenty times the energy contained in the mass of a proton. These high-energy electrons were shot at a vessel containing liquid hydrogen, where some of them hit the protons contained therein.

If only so-called *elastic* collisions were considered, everything behaved as usual. In these collisions, electrons and protons behaved like perfectly elastic rubber balls, which only exchange kinetic energy and momentum. No energy was used to create internal proton excitations or new particles. So the protons as a whole reacted to the collisions without changing internally. Such experiments had already been carried out in the past, albeit at lower energies. There the proton had always behaved like a soft, diffuse charge cloud whose radius was just under one fermi (1 fm $= 10^{-15}$ m) without any substructure. The new higher energy elastic collisions at SLAC confirmed this result.

But such elastic collisions could not reveal a possible quark structure, because the protons reacted as a whole. So it was important to look at *inelastic* collisions, where so much energy is transferred to the proton that it is either converted into an excited quantum state – for example, the proton becomes a Δ^+ baryon – or completely new particles are formed – in which case we also speak of *deep inelastic* collisions. The virtual photon, which transfers the large amount of energy from the electron to the proton, has a correspondingly short wavelength and can resolve details inside the proton.

When such deep-inelastic electron–proton collisions were investigated in 1968 at SLAC, the results were astonishing. Surprisingly, quite a few electrons were found to be significantly deflected from their path by the protons. If protons really were only diffuse charge clouds, there should have been far fewer significantly deflected electrons – as if a bullet had been shot through a block of butter. A diffuse proton could not have offered the electrons enough resistance to create a substantial deflection. However, the experimental data seemed to be saying that the electrons had hit "hard" point-like objects inside the protons and been thrown off course by them. It was as though the butter contained little stones, so to speak, which could throw the bullets off course.

More precise analyses also showed that the inner structure of the protons did not affect the electrons as it might have done in the most general case. The scattering data showed certain mathematical peculiarities (so-called scaling properties) that begged an explanation – so here was another mystery!

Bjorken Scaling – It Worked, but Hardly Anyone Understood It

The American physicist James Bjorken was working at SLAC when these results became known in 1968. At the age of 34, Bjorken was significantly younger than Feynman, who was already in his fiftieth year. As an aspiring theoretical physicist, Bjorken was well versed in the modern mathematical methods of quantum field theory and group theory. He was also familiar with the so-called current algebra which Murray Gell-Mann had developed. This was an abstract mathematical method for calculating collision processes of baryons and mesons under certain assumptions. With this current algebra, Bjorken was able to explain mathematically why the scattering data showed the observed mathematical characteristics – today this is also referred to as *Bjorken scaling*. However, for most physicists, Bjorken's derivation was far too abstract – they didn't understand what he was doing, even though the results looked promising. So what did Bjorken's results mean?

Feynman's Parton Model of the Proton

This was exactly the kind of situation where a physicist like Richard Feynman could be useful. For here was someone who could explain in simple terms what people were actually seeing in experiments. Fortunately, Feynman was still interested in the ongoing collision experiments and was trying to make sense of the data from them. Little by little, the following idea began to take shape:

Suppose there were actually some point-like objects inside a proton. Feynman called them *partons* to be on the safe side, because he did not yet want to commit himself to Gell-Mann's quarks or any other special particles, but rather wanted to formulate his idea in general terms. What would happen if very high-energy particles such as the electrons at SLAC collided with the partons in a proton?

Feynman imagined what a stationary proton would look like from the perspective of an electron approaching almost at the speed of light. From the electron's point of view, the proton comes racing towards it almost at the speed of light. At this high speed, the effects of special relativity play a decisive role. The so-called time dilation, for example, causes all processes inside the proton to slow down considerably from the electron's point of view. If there are any partons in it, their lateral motions will be almost frozen. In addition, the Lorentz contraction causes the proton to flatten in the direction of motion. Thus for a very high-energy electron, the proton looks like a flat pancake with the partons embedded almost motionlessly in it like raisins. In fact, it is the raisins alone that make up the proton, so the electron sees a frozen flat cloud of point-like partons coming towards it.

With this picture in mind, Feynman visited SLAC in August 1968 to find out about the latest experiments on site. As SLAC was only a few hours away from Caltech by car, such a visit was an obvious choice. Feynman was also able to combine this with a visit to his sister Joan, who worked at NASA's Ames Research Center at the time and lived not far from SLAC.[1]

Unfortunately, Bjorken himself was not on site at the time, but the other physicists showed Feynman the latest data and Bjorken's conclusions. However, nobody could explain to Feynman how Bjorken had come up with it.

Feynman suspected that he could achieve the same results with his parton model. So he sat down and calculated overnight what his model had to

[1]See Christopher Riley: *Joan Feynman: From auroras to anthropology*, auf http://findingada.com/.

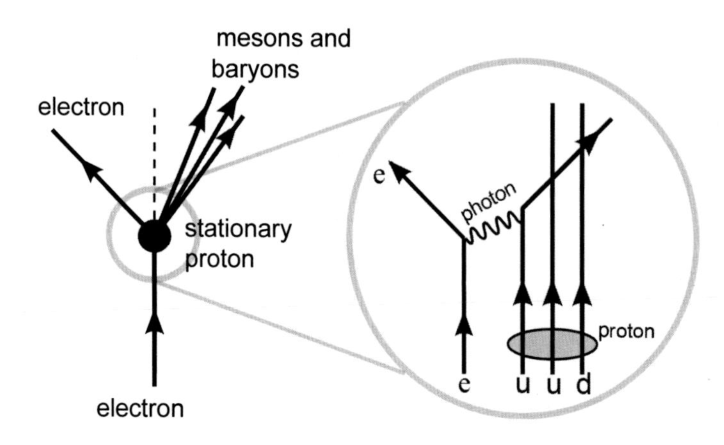

Fig. 6.3 Deep inelastic collision of an electron with a stationary proton. The electron emits a virtual photon which collides with one of the partons (here, an up quark)

say about the deep inelastic scattering of electrons on protons. The result was very simple: the virtual photon that is emitted by the incoming electron hits a single parton inside the proton and catapults it out in a certain direction (Fig. 6.3). Since the proton is almost frozen from the electron's point of view, the parton has no time at all to interact with the rest of the proton at the moment of the collision, so it behaves at first almost like a free particle. Only a little later, when the parton wants to leave the proton, will the interaction with the other partons in the proton become noticeable and a jet of different particles will be created from the energy of the parton.

So in deep inelastic processes, the high-energy electrons are not at all deflected by the whole proton, but only by a single point-like parton. If we look at the proton from the electron's point of view again, the proton and its constituent partons rush towards the electron with a high momentum. Each individual parton can now have a fraction x of the total momentum of the proton with some probability. These probabilities, which depend only on the dimensionless parameter x, are the only effect that the inner structure of the proton can have on the deflection of the electron in this scattering process. This was exactly what Bjorken had found using his current algebra.

When Bjorken arrived at SLAC, Feynman presented his ideas to him. Bjorken, who was used to abstract mathematics, remembers in *Most of the Good Stuff* how Feynman expressed himself: it was an easy, seductive language that everyone could understand.

It was no wonder therefore that Feynman's parton model quickly spread among physicists for whom Bjorken's style was too abstract. Finally, everyone could understand what was really behind Bjorken's findings. The results of the deep inelastic collisions proved that the proton really did consist of smaller constituents. Feynman had once again explained physics to the world.

From Partons to Quantum Chromodynamics

For Gell-Mann – the inventor of the term "quark" – the relatively simple parton model was once again one of Feynman's typical tricks for making physics accessible to people who could not understand more complex theories. So was Feynman just hijacking his quark theory? Gell-Mann was also ruffled by the word *parton*, which for him was an ugly cross between Latin and Greek.

In contrast to Gell-Mann's concerns, however, Feynman had just taken an unbiased approach and simply wanted to find out in general what the available experimental data meant. Whether the now proven partons were identical to Gell-Mann's quarks still had to be proven by further experiments. For example, did they really carry the weird fractional charges?

Over the following years, many more deep inelastic scattering experiments were carried out at SLAC and elsewhere to determine the exact properties of the partons. Gradually, it became clear that they actually had exactly the characteristics of the quarks. For example, it could be shown that a proton actually contains two up quarks and one down quark, which together determine the charge and spin of the proton. In addition to these so-called valence quarks, other components of the proton could be detected: a sea of constantly emerging and disappearing virtual quark–antiquark pairs (see Sect. 3.1), together with electrically neutral particles which were given the name *gluons* and which carried about half of the proton's momentum. So if we look closely, the proton turns out to be an extremely complex quantum object. Feynman's parton model allowed for this complexity, while Gell-Mann's quark model was limited in this respect.

After the existence of quarks had finally become more widely accepted, a rapid development started. In 1971, the German physicist Harald Fritzsch realized together with Gell-Mann how the problem with the Pauli principle could be solved for the Ω^- baryon and other particles. They suggested that each quark had an additional property besides its flavor, which they called *color charge* or just *color* – although, of course, it has nothing to do with real colors. This color charge can take the values *red, green*, and *blue*, while antiquarks can have the corresponding colors *antired, antigreen*, and *antiblue*. The three otherwise identical quarks in the Ω^- baryon then differ by their color charge and thus assume different quantum states – the Pauli principle was thereby saved!

A year later, Fritzsch and Gell-Mann discovered that the colors of the quarks could be regarded as the fundamental charges of the strong interaction. Just as photons react to electrical charges and can be emitted and absorbed by them, so gluons react to the color charges of quarks. On this basis, the two physicists were able to formulate a theory of the strong interaction analogous to QED, which they called *quantum chromodynamics* (QCD). However, there is an important difference between QCD and QED: gluons carry color charges themselves, while photons are electrically neutral. This considerably complicates the situation. Nevertheless, QCD has proven itself in all experiments to date and is considered today to be the correct fundamental theory of the strong interaction.

Quarks – Locked up Forever and yet Sometimes Almost Free

Quarks and gluons do not exist as free particles – they are locked up inside baryons or mesons. Free color charges, like single quarks, cannot exist. They must be neutralized in some way with respect to their surroundings. Only particles that do not carry an overall color charge, i.e., "color-neutral" particles, can exist as free objects. Therefore, quarks must always combine with other quarks or antiquarks to form mesons or baryons, in such a way that their colors add up to white, so to speak. This can be done in two ways: either by combining a color and its anti-color (e.g., red and antired) or by combining three different colors or anti-colors (either red, green and blue or antired, antigreen and antiblue).

Many people have tried to derive this result, known as *confinement*, directly from QCD, but this has not yet been achieved in a truly satisfactory way. In visual terms, confinement is caused by the fact that gluons themselves carry color charges and can therefore interact directly with each other. If two quarks move apart, the field lines of the strong interaction between them attract one to the other ever more strongly and end up forming a kind of tube of interacting gluons between the quarks. If the distance between the quarks increases further, this elongating tube causes an ever stronger attraction between the quarks – something like a tightly stretched rubber band, which is increasingly difficult to pull apart the more stretched it is. At some point the energy stored in the tube is so great that it is sufficient to form new quarks and antiquarks. The tube snaps and these new quarks and antiquarks settle at the fractures, creating new mesons and baryons.

The mutual interaction of the gluons has an additional effect. The color charge of the cloud of virtual gluons surrounding the quark increases with increasing distance, and it is precisely for this reason that it must be neutralized to the outside by the color charges of other quarks. On the other hand, very close to a quark, the strength of its color charge seems to decrease. This is known as *asymptotic freedom*, which was discovered in 1972 and 1973 by Gerard't Hooft, David Gross, David Politzer, and Frank Wilczek.

The effect with color charges is therefore exactly the opposite to what happens with electric charges, since the latter grow stronger closer to the charge due to vacuum polarization, as we saw in Sect. 3.3. In deep-inelastic electron–proton collisions, the quark hit by the photon moves very fast, so only its immediate surroundings can react to it at first. At this short range, however, the color charge of the quark is still quite small, so initially it moves almost freely – just as Feynman had assumed in his parton model. Only a little later, when the quark tries to escape from the proton, does the stronger color charge have an effect: the gluon tube connecting it to the other quarks in the proton forms and finally snaps, producing further quarks and antiquarks.

Since the color charge of quarks increases with growing distance, Feynman diagrams can only be used at very small distances in QCD. In the case of deep inelastic collisions, this is only the case in the initial phase when the quark was just hit. Feynman made corresponding calculations, in the mid to late 1970s, with Richard Field, who was a young postdoctoral student at Caltech at the time. Feynman was enthusiastic about the development of QCD and tried to calculate everything that could somehow be calculated using his diagrams with QCD – just as he had done in the past with QED.

The formation of new mesons when the gluon tube snaps cannot be calculated with Feynman diagrams, since greater distances and thus larger color charges are involved there. The same applies to the phenomenon of confinement. With large color charges, it is no longer justified to limit oneself to the simplest diagrams. Let us remember why. Each vertex in a diagram leads to a charge factor in the corresponding mathematical expression, so that complex diagrams contain many charge factors. In QED, the elementary electric charge is small, so complex diagrams make only minor contributions. In QCD, on the other hand, the color charge is large at greater distances, so complex diagrams cannot be omitted. One gets quickly lost in the growing number of diagrams, and it is also questionable whether they still deliver any meaningful results overall.

One way out is to use computer calculations within the framework of the so-called lattice QCD, which is not based on Feynman diagrams. Feynman also became involved in such calculations. They provide clear indications that confinement should exist, but they still do not really provide a proof. Feynman himself also tried to find such a proof with the help of his path integrals, but had no conclusive success. A convincing solution to this problem has yet to be found.

The Standard Model Emerges

The successful formulation of QCD was followed by a whole series of other important discoveries and theoretical developments. New heavier quarks and leptons were discovered and it was revealed how electromagnetic and weak interactions could be combined into a unified theory. By the end of the 1970s, the result was the so-called Standard Model of particle physics, which we already encountered in Sect. 4.3. This jewel of theoretical physics describes all known matter consisting of quarks and leptons, as well as the weak, electromagnetic, and strong interactions between these particles in the framework of a common quantum field theory. Only gravity is left out, as there is still no consistent quantum theory for it. In all experiments carried out to date, the Standard Model has always been corroborated to high accuracy.

Nevertheless, it is clear that the Standard Model cannot be a truly fundamental theory of nature. Not only does it contain too many free parameters, such as particle masses and charges, which cannot be explained within the theory, but much more seriously, gravity is not contained in the model. Furthermore, with dark matter and dark energy, we have also discovered clear indications that there are forms of matter in the universe that cannot consist of quarks and leptons. So there are plenty of open questions that the Standard Model cannot answer.

Unfortunately, Feynman, whose work helped to build so much of the foundations of the Standard Model, was less and less able to participate in its formulation. Since he was first diagnosed with cancer in 1978, he could no longer participate in frontline research as he had done previously. Even someone like Feynman couldn't go on forever. But he still had enough energy to turn to an old love that had been with him since his time in Los Alamos: computers and some of their more wonderful features.

6.2 Computers

Feynman had his first intense contact with computers as early as 1944, when he was working on the Manhattan project in Los Alamos. There he led a group that was supposed to use simple calculating machines to figure out what would happen when the bomb exploded, and how much energy would be released. Today one can solve such a task approximately with any home computer. But nothing like this existed at that time. What they had were simple electronic machines from IBM that could multiply, add, or sort numbers – depending on the type of machine. The numbers had to be encoded on punch cards and inserted into the machines. So the IBM group was constantly juggling stacks of punch cards back and forth between the different machines. How could the calculations be carried out effectively under such conditions? How should the punch cards be transferred between the machines and how should one organize the individual tasks in the group to process as many calculations in parallel as possible? That was the challenge Feynman and his group had to face. Over time, they developed increasingly sophisticated ways to parallelize the calculations.

Feynman tells an interesting anecdote in *Los Alamos from Below*, which shows how much he was already fascinated by computers back then. Working on the project there was a rather clever guy named Stanley Frankel, who ordered the calculating machines from IBM and built up the team at the beginning. But he began to suffer from the computer disease that anybody who works with computers now knows about, according to Feynman: you start playing with the computers. "They are so wonderful", Feynman says: "You have these switches – if it's an even number you do this, if it's an odd number you do that – and pretty soon you can do more and more elaborate things if you are clever enough, on one machine." So Frankel began to neglect the team and spent hours sitting at one of the computers, trying to get it to calculate the arc tangent of various angles automatically, for example. This was absolutely useless, because one already had long tables giving the values of this mathematical function. But Feynman could understand Frankel well: it was simply a great delight to see how much you could do. However, things could not go on like this, so Feynman finally took over the group and made sure that he did not succumb to the same disease – which was certainly not always easy for him.

In the 1980s, at over sixty years of age, Feynman was able to pursue all his computer interests to his heart's content. He was probably also motivated by the fact that his son Carl was then studying computer science. In any case, from 1981 Feynman gave a series of lectures at the Caltech in collaboration with a group of leading computer scientists, and these were also published in book form in 1996 under the title *Feynman Lectures on Computation*. In addition to the basics, he dealt in particular with the areas where computers and physics meet, including for example the thermodynamics of computers and the potential of quantum computers. Again and again he asked himself what computers could do in principle and where the insuperable physical limits of computers might lie. Feynman sought contact with many pioneers of computer science, inviting them to give guest lectures and thus learning much about this discipline, which was quite new to him.

For Feynman, however, "learning" was something different than for most people. It meant discussing topics intensively, playing around with them, and working on them as independently as possible, on the basis of just a few hints. Once he had found something interesting, he would walk around trying to explain it to anyone who was willing to listen. Only later did he occasionally check whether others had already achieved the same results. It was not important to him whether he created something new. He just wanted to solve problems and understand contexts. If the leading experts in computer science were ahead of him, he was happy that he had also achieved the same result in his own way – not bad for an amateur, he would think!

Parallel Computers and the Connection Machine

In addition to his lectures and theoretical investigations, Feynman also dealt with computers on a quite practical level. He was particularly interested in the question of how a task could be distributed across several processors and executed in parallel, as this was just the problem he had been dealing with in Los Alamos almost forty years previously. A project started by the young computer expert Danny Hillis, who worked at MIT at that time, was particularly ambitious in this respect. Hillis wanted to build a new type of parallel computer whose performance would be so powerful, with up to a million small processors, that it would be capable of human thinking – in other words, a truly intelligent machine. The basic idea was quite plausible. Our brain works very much in parallel, with its innumerable neuronal connections. So a large parallel computer is needed if we hope to emulate the brain's abilities.

From today's standpoint, this project was certainly a bit naive. But the idea of massive parallelization was certainly forward-looking. When writing these lines in June 2018, the fastest supercomputer *Sunway TaihuLight* in China uses 40,960 processors, each with 260 processing cores, so this supercomputer has a total of over ten million (to be precise, 10,649,600) CPU cores.

But even our modern supercomputers are only just beginning to be capable of artificial intelligence, although there has been impressive progress in this area recently. Back in the 1980s, it was impossible even to estimate what enormous computing power and sophisticated algorithms would be required to teach a computer abilities such as image and speech recognition, or even human thinking. And it was not known what it meant to connect a million processors to each other in such a way that they could exchange messages effectively – so in the end they installed, not a million, but only 64,000 processors, which was still an impressive number at the time.

Feynman's son Carl, who was studying computer science at MIT at the time, helped Hillis with his project. Finally, Hillis founded a company in 1983 with the promising name *Thinking Machines Corporation* to put his project into action and build the parallel computer he called *Connection Machine*. The motto of the young company made clear what Hillis was all about: "We are building a machine that is proud of us!"

Hillis gathered a group of outstanding computer experts around him. Carl also introduced Hillis to his father, and Hillis told him about his plans. Feynman, who was always interested in his son's activities, replied that this was positively the dopiest idea he had ever heard of. This was by no means meant in a derogatory way, because of course Feynman was interested to find out whether such a thing was actually possible. So he agreed to work as a consultant for the new company.

When Feynman arrived at the company's office in Boston and asked how he could help, he was told that he should only advise the company on how to apply parallel computers to scientific problems. But it was too early for that, because the computer didn't exist yet – so he would just stand around wasting his time. That was not Feynman's cup of tea, so he insisted: "Give me something real to do!"

The critical component of the parallel computer was the so-called router, which had to ensure communication between the thousands of processors. So they asked Feynman to analyze whether the planned design for the router would work. That was not an easy task, because with so many processors it was impossible to connect every processor directly to every other processor – twenty direct connections per processor was the maximum. So the router had to find the best route for each message in the complex traffic of other messages between the processors.

That was just the right thing for Feynman, and he literally began to simulate with paper and pencil how the router worked. Feynman loved details – he was convinced that this was the only way to understand what was really going on. The result of his analysis was a collection of several differential equations that described the mean behavior of the system statistically. For physicists such an approach was normal, but for the computer experts involved in the project, this approach seemed a bit strange. Their analysis had shown that seven small buffers were needed per processor to buffer a result until a path to the target processor became available. According to Feynman's analysis, five buffers would suffice.

Could Feynman's unconventional analysis be trusted? People were suspicious and decided to play it safe and equip the processor chips with seven buffers. But then it turned out that such chips were difficult to produce. So in the end they took the risk and built chips with only five buffers – and behold, the computer actually worked with this slimmed-down design. Feynman was right!

Feynman did not limit himself to solving this router problem, but developed a multitude of activities. For example, he contributed an algorithm to calculate logarithms very efficiently on a parallel computer. Feynman had already found this algorithm in Los Alamos and was very proud of it.

In Los Alamos Feynman had also learned how to organize a larger group so that they could work together successfully. The young company now benefited from this knowledge. As a result, individual teams were formed with team leaders and a regular seminar was held, to which many top-class guest speakers were invited. From them, Feynman and the other employees learned a lot about possible applications of parallel computers.

Feynman invited a friend from Caltech to the first seminar: the American physicist and neuroscientist John Hopfield, who gave a lecture on neural networks. Such networks are based on an analogy with the brain and are very well suited to learning and recognizing patterns, such as images of letters. Since all neurons work almost in parallel in a neural network, a parallel computer is ideal for simulating this. Feynman was thrilled and worked out the details of how each processor could be used as a neuron.

Massively parallel computers are suitable for a variety of other problems. As with neural networks, applications often involve simulating something: the progress of evolution, the interior of the Earth, the development of stars, the folding of proteins, and much more. The behavior of quantum chromodynamics (QCD) can also be simulated by approximating space and time through a lattice of points and calculating the mutual interaction of the quantum fields at these points. Feynman himself wrote a corresponding computer program in the only programming language he knew: BASIC.

Cellular Automata: Simple Rules Create Complex Behavior

The very first program to run on the Connection Machine in April 1985 was the *Game of Life*, which was invented by the English mathematician John Horton Conway in 1970. Actually, it is not a game, but a two-dimensional *cellular automaton*, i.e., an artificial world that can be imagined as a very large piece of squared paper. Each box – also called a cell – can be either black or white. In other words, the cell can be alive or dead (hence the name).

You start with some pattern of living and dead cells in the grid world, then calculate what the next generation of cells will look like, then the generation after that, and so on. The rules for determining the next generation are very simple. Each cell has eight neighbors, including the cells diagonally adjacent at the corners. Whether a cell in the next generation is alive or dead depends only on its current state and the current state of its eight directly neighboring cells. If the cell itself is dead and has exactly three living neighboring cells, they bring it to life. A living cell that has only one or no living neighbor dies of loneliness. With more than three living neighbors, on the other hand, it dies of overpopulation. So it only stays alive with two or three living neighbors, and exactly three living neighbors can even bring a dead cell to life.

You can even play this simple game on a piece of paper. Of course, a computer is much better for this, and a parallel computer, where each processor takes care of one cell, is particularly suitable. So it was no coincidence that Conway's Game of Life was chosen to test the Connection Machine.

Depending on the starting pattern, the simple rules in the game of life lead to very different, constantly changing patterns that can become very complex and sometimes even develop chaotically. Some patterns remain unchanged, others grow or shrink, some even oscillate or move across the grid (Fig. 6.4). You can even find patterns that simulate computers. Conway's Game of Life, which is simulated on a computer, can therefore be something like a computer itself. And since a computer can simulate a great deal, the game of life or, more generally, a cellular automaton can also simulate a great deal. It's hard to believe what a few simple rules can produce on a grid!

With the young British–American physicist and mathematician Stephen Wolfram in the consulting team, they had someone who was very familiar with cellular automata and their applications. Wolfram had obtained his doctorate at Caltech University and subsequently worked on the development of a computer algebra system that could transform and calculate with terms like a mathematician. This would later become the world-famous program package *Mathematica*, used in so many areas of science, engineering, and mathematics.

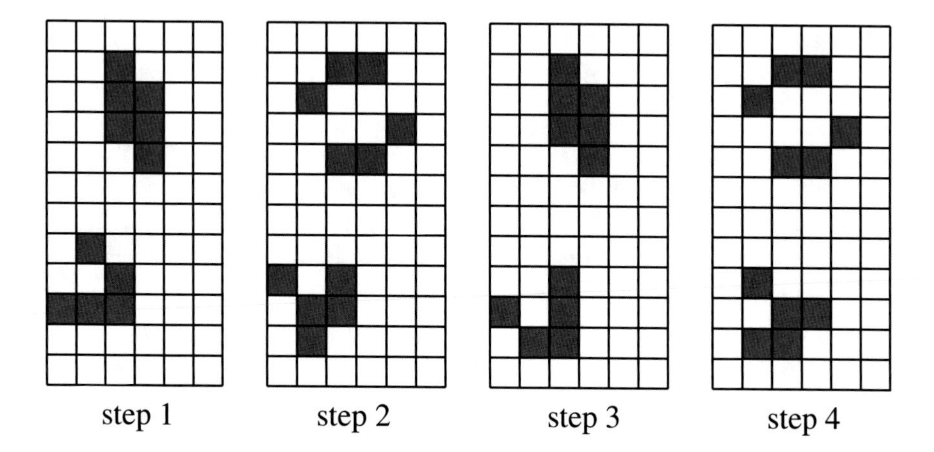

Fig. 6.4 Four steps in Conway's Game of Life. Above you see a periodically oscillating object called toad. Below is a so-called glider, which moves across the grid further and further to the right below

Stephen Wolfram, who is considered highly intelligent and somewhat headstrong, was over forty years younger than Feynman. They knew each other from Caltech. In *A Short Talk about Richard Feynman*, Wolfram recalls how in the summer of 1985 he showed Feynman a particularly interesting cellular automaton that he himself had found two years earlier. This machine, called *rule 30*, is even simpler than the Game of Life, because it doesn't need a two-dimensional grid as a playing field, but only a one-dimensional row of cells. So each cell has only two neighbors. Whether a cell of the next generation is reborn, remains alive, or dies is determined according to the following simple scheme:

Old state (cell and both neighbors)	111	110	101	100	011	010	001	000
New state for center cell	0	0	0	1	1	1	1	0

Here, a 1 represents a living cell and a 0 a dead cell. The combination 001 in the second-to-last column therefore means that the center cell itself is dead (middle 0), as is the left neighbor, while the right neighbor is alive. In this case, the cell is brought to life (new state 1). If the bottom line 00 011 110 is interpreted as a binary number, it represents the number $2+4+8+16=30$; hence the name *rule* 30.

If we look closely at the rule, we discover an asymmetry: the combination 110 causes the center cell to die, while combination 011 causes the cell to live. So if the cell and its left neighbor are alive while the right neighbor is

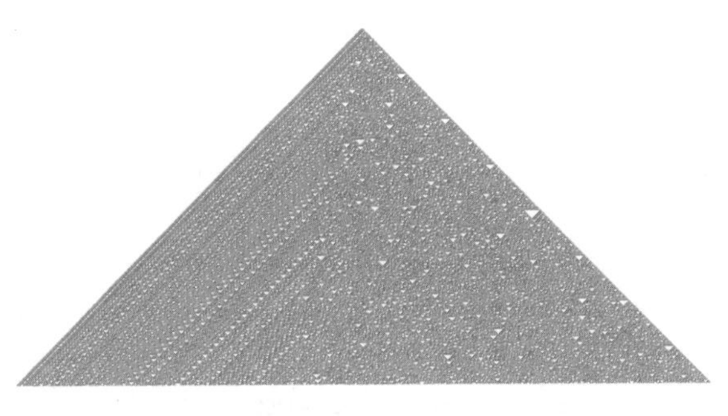

Fig. 6.5 The first 500 generations created from a single cell according to rule 30
Source: https://commons.wikimedia.org/wiki/File:Rule30-first-500-rows.png

dead, the cell dies. If we swap the two neighbors, it stays alive. If the cell itself is already dead, the situation is symmetrical, because both combinations 100 and 001 bring the cell to life.

Interestingly, even this simple set of rules leads to very complex behavior due to its asymmetry, as shown in Fig. 6.5. If we start with a single living cell, the line of living and dead cells expands with each step. If the individual time steps are arranged one below the other, a kind of triangle is formed, which shows a regular stripe pattern on the left. If we go further to the right, this regularity suddenly gets lost and becomes chaotic, whereupon white triangles of different sizes appear again and again.

Feynman and Wolfram puzzled over how the simple rule 30 could create such a complex structure. Feynman, together with his son Carl, tried to gain some kind of intuition for the workings of the rule using all the tools they had at their disposal. How did the borderline between order and chaos develop, and what caused order to turn into chaos? Eventually, Feynman gave up – this seemingly simple automaton had actually pushed him to his limits! Perhaps there really is no other way to understand the behavior of the rule than simply to let it run and see what happens – as suggested by Wolfram. There may not be a deeper reason than the rule itself, in which case, Feynman had no chance of finding one.

In *A Short Talk about Richard Feynman*, Wolfram describes how Feynman conspiratorially took him aside and asked how he knew rule 30 would do all this crazy stuff. "You know me," Wolfram replied. "I didn't. I just had a computer try all the possible rules. And I found it." Feynman was relieved: "Now I feel much better. I was worried you had some way to figure it out."

Depending on the applied rules, cellular automata show such diverse behavior that even physical systems can be simulated with them. Using a suitable cellular automaton on the Connection Machine, Wolfram created a wonderful film that seemed to show a two-dimensional turbulent fluid. The impression of a liquid is created by looking at many cells, so that they merge into a continuum from a distance. Even a real liquid is not continuous, but consists of many individual molecules that move and collide with each other. Macroscopically, this microscopic structure then merges into a continuous entity which can be described by differential equations. Microscopically, however, these differential equations between the individual molecules do not apply – rather, they arise when the microscopic behavior of these molecules with all their ongoing movements and collisions is smoothed out and statistically averaged. It's the same with a cellular automaton: only the average behavior of many cells creates the impression of a continuous fluid.

Simulating Nature – Is the World a Computer?

Could it be that nature itself is not continuous? Are space and time on their lowest level perhaps similar to a cellular automaton, so that the known laws of nature only emerge as an average of the microscopic behavior of the individual cells? What rules would these cells have to obey in order to create the world as we know it?

Such thoughts had been on Feynman's mind for quite some time. He began to ask himself whether the laws of nature could in general be simulated on a computer, and what such a computer should look like. What can we learn about nature if we simulate it on a computer? Is the universe itself perhaps nothing more than a gigantic computer, as suggested by the German engineer and builder of the first programmable computer, Konrad Zuse, in 1969, and published in his book *Calculating Space (Rechnender Raum)*?

In 1981, Feynman gave a remarkable keynote speech at a conference on the *Physics of Computation* at MIT, entitled *Simulating Physics with Computers*. Despite his initial remark that he had no idea what a keynote speech was, he managed to give an excellent keynote speech. In this, Feynman discussed the question of whether there could be a universal computer capable of simulating the laws of physics with absolute precision. His aim was not to approximate the laws of nature using some numerical algorithms. Rather, the computer should do exactly the same as nature does. Was that possible?

Feynman specifies two conditions that the computer must meet: its components must be *locally interconnected*, i.e., information should only flow between adjacent components – similar to a cellular automaton; and the number of its components should only increase in proportion to the volume of space to be simulated physically. The same applies to the required computing time – it should also increase in proportion to the simulated time span. So we don't want to have a computer of unlimited size, inside which long wires connect remote elements and which takes arbitrarily long times to carry out its simulations.

Now such a spatially limited computer can only execute a limited number of logical operations within a certain period of time. If this computer is to accurately represent the physics taking place in a limited spatial volume and over a certain time span, it must be possible to analyze this physics with a limited number of logical operations. This is not the case in classical physics, since this is based on continuous quantities such as the time and space coordinates of particles and the strength of force fields. This does not change in quantum theory, which also uses continuous time and position coordinates and continuous quantum amplitudes. Such a continuous quantity can contain an unlimited amount of information, captured by an infinite number of decimal places after the decimal point. A computer, on the other hand, can only work with a finite number of decimal places.

Must we conclude that physics cannot be exactly simulated with a computer? This would be the case with physics as we understand it today, but there is another solution: the known physics with its continuous quantities may only be an approximation. It could be that real physics is basically discrete. Space could consist of individual points or cells, and time could progress in jumps. The impression of continuous quantities would emerge by smearing the microscopic details. The same was true for the simulation of fluid flows with a cellular automaton. Feynman liked the idea that space and time could be discrete: otherwise, a finite volume would be able to absorb an infinite amount of information. That didn't seem plausible to him.

The idea of constructing space and time from discrete elements like cells leads to some difficulties. Can we then ensure that all directions in space are equivalent? Is the equivalence of different reference systems in relative motion still guaranteed? These questions are not easy to answer. On the other hand, considerations of quantum gravity clearly indicate that deep down nature is not continuous. There is much to suggest that the Planck length and Planck time are the smallest space and time intervals that still make sense physically.

The Symmetry of Time: Reversible Computation

What about the other characteristics of nature? Do they match the behavior of a computer? One potential problem could be that the fundamental laws of nature are reversible, i.e., time-symmetrical (except for a small asymmetry in the weak interaction, which we can safely ignore here). If we look, for example, at the collision of two molecules or at the motion of a planet in the reverse time direction, we will see another physically possible process. We won't be able to tell whether we're watching the corresponding movie forwards or backwards. This even applies to electromagnetic fields, as we have seen in Sect. 2.2. The difference between the future and the past only emerges in systems with very many particles, which evolve from a statistically unlikely state to a more probable state.

To enable a computer to exactly simulate a time-symmetric process, the corresponding calculation process must also be reversible in time. So if we look at the calculation process backwards in time, we should see another possible calculation process. This is not the case with today's most common computer architectures. If, for example, two bits are linked via a logical AND, a zero results as soon as at least one of the two input bits is equal to zero. The two initial bits can no longer be reconstructed from the result.

However, one can change the computer architecture by taking care not to lose any information during the calculation process. For example, for the AND, we would have to use additional bits to ensure that the two input bits can be recovered from the calculation result. This is not particularly efficient technically, but it can be made to work! Reversible computing is thus possible in principle, as shown by the work of Rolf Landauer, Charles Bennett, and Edward Fredkin, among others. Feynman had learned a lot about reversible computing from Charles Bennett, and he already knew Edward Fredkin from Caltech in 1974. There they had made an agreement at the time: Feynman would teach Fredkin quantum mechanics, while Fredkin would explain the basics of computer science. However, it was difficult to teach Feynman something in the usual sense, because he always wanted to find everything out by himself. He only ever wanted a few hints and was soon upset if he was told too much – this would deprive him of the pleasure of discovering it for himself. Not fame and honor, but "the pleasure of finding things out" was the driving force in his life. It was not for nothing that this pleasure later became the title of one of his books.

Simulating Space and Time

The time symmetry of the laws of nature should not therefore be a problem for an exact computer simulation. What about the simulation of time itself? If we assume that time does not elapse continuously but in discrete steps, then a cellular automaton, for example, can simulate this well, by taking steps from state to state, as it does. But actually, according to Feynman, the computer just imitates time. And time could be more complicated, so mere imitation might not be enough.

Let us look at the example of a one-dimensional discrete space, as in the case of the cellular automaton used for "rule 30" above. In Fig. 6.5, we plotted the individual time steps downwards, so that a two-dimensional space-time grid is created in which each cell represents a certain cell at a certain time. For rule 30, the state of a cell at a given time is determined by its own state and the state of its two neighboring cells at the previous time. But nature could also work in a much more general way. It may also be necessary to take into account other cells that are further away, both in space and time, when determining the state of each cell. The cells could even be in the future, i.e., the future could also influence the past. This may seem strange, but we have already seen several times in this book that such a concept can make sense, for example, with advanced electromagnetic waves, or again with positrons, which we can understand as electrons traveling backwards in time. "That could be the way physics works", writes Feynman in *Simulating Physics with Computers*.

Can a computer calculate something like that? Can it determine the state of each cell in space and time when many or even all cells influence each other across both spatial and temporal separations? The result is a complicated self-consistency problem that is not always solvable – not even for a computer.

In classical physics, such complicated things should never occur. Classical physics is causal and reversible, i.e., the immediate past determines the future and vice versa. At the same time, it is local, i.e., if we assume a discrete structure of space and time, only the neighboring cells determine the next state of a cell. "We have no difficulty, in principle, apparently, with that", says Feynman.

Quantum Mechanics and Computers

However, nature does not work according to the rules of classical physics, but rather according to the laws of quantum mechanics. An exact simulation of nature must therefore reflect the quantum mechanical phenomena in every detail.

In quantum mechanics, the results of an experiment can generally no longer be predicted. We cannot calculate through which hole an electron will pass in the double-slit experiment and where it will hit the detector screen. All we can calculate are the corresponding probabilities.

Feynman admits in *Simulating Physics with Computers* that we have always had a great deal of difficulty understanding the world view suggested by quantum mechanics. At least Feynman himself did, as he writes there, and despite his age, he felt that he still hadn't got to the point where it was obvious to him. "I still get nervous with it", he writes. For every new idea, it takes a generation or two until it becomes obvious that there's no real problem with it, and quantum mechanics has not yet reached that point. As he said: "I'm not sure there's no real problem." And this motivated him to ask all these questions about computers and whether they could simulate nature – was it possible to learn something from this about the mysteries of quantum mechanics?

Quantum mechanics makes assertions about probabilities, so we have to ask ourselves how a computer can simulate probabilities. In physics, these probabilities are usually simply calculated. However, probabilities are continuous quantities that can only be represented approximately by a finite number of decimal places in a computer. If we want an *exact* simulation of nature, that would not be enough.

There is another problem. In order to describe, e.g., liquid helium quantum mechanically, we have to specify a probability for each of the countless atomic configurations. We have to say how likely it is to find an atom at location A, another atom at location B, a third atom at location C, and so on, and we have to do this for all sorts of combinations of locations. If we again assume that space is divided into discrete cells – say N of them – and that we are dealing with R helium atoms, there are approximately N^R possible spatial distributions of the atoms (configurations) in the cells, and we would have to calculate all their probabilities. Since for a bottle of liquid helium both N and R are very large numbers, N^R is a gigantic number – much larger than the number of atoms in the universe. Such a large number of probabilities could never be calculated or stored, and that is exactly the problem with any quantum mechanical calculation. Even in simple atoms with only a few electrons, the wave function of the electrons could at best only be determined approximately.

The problem lies not only in the pure size of the number N^R of configurations, but also in the fact that it grows exponentially with the system size. If we want to calculate the probabilities of all configurations for two liters of liquid helium instead of one liter, we do not need a computer twice as large. If the number of configurations is N^R for one liter, it is $(2N)^{2R} = 4^R \cdot N^R \cdot N^R$ for two liters, so the number of configurations has increased by a huge factor of $4^R \cdot N^R$. The computer would have to be correspondingly larger and faster to cope with the calculation. But earlier we insisted that the number of computer components should only increase in proportion to the volume it simulates.

Thus, calculating probabilities or wave functions will not help us to achieve an exact simulation that fulfils our conditions. But there is another way: if nature behaves in a random way and we can only consider the corresponding probabilities anyway, it suffices for this to apply also to our computer. The computer should behave just as randomly as nature itself. If we start such a probabilistic computer in a certain initial state, it will end up in different final states with certain probabilities. Exactly the same happens in nature: if we shoot an electron through a narrow double slit, it will hit the detector screen with certain probabilities at different locations. The probabilistic computer is now supposed to simulate nature in the sense that the probabilities are the same.

It is not therefore necessary for the computer to behave exactly like nature. "You see, nature's unpredictable; how do you expect to predict it with a computer? You can't!" stated Feynman. However, it should be possible to repeat the real experiment and the corresponding simulation on the probabilistic computer a large number of times and compare the frequency of a given final state. For an exact simulation, the more often you repeat experiment and simulation, the better they must match statistically.

Is it possible to simulate the quantum mechanical behavior of nature exactly with such a classical probabilistic computer? To answer this question, we need to be more specific about how our computer should work. We want it to operate *locally*, similarly to the cellular automata: the probabilities for the individual states of a cell in the next time step should therefore only depend on the current state of the cell and its neighboring cells, and not on the states of cells further away. In order to determine the probabilities in a certain region for the next time step, we can thus simply ignore what is currently happening in more distant regions.

Simulating the Electron Spin

As a typical example from the quantum world, let's go back to the spin of an electron and take a closer look at its properties. We already know that the electron behaves rather like a tiny magnetic compass needle because of its spin. Julian Schwinger had been the first to calculate the magnetic strength of this compass needle – the so-called magnetic moment of the electron – using QED, in 1947.

The spin always causes the electron to be drawn either into or against the direction of the magnetic field lines in an inhomogeneous magnetic field. If a beam of electrons flies through an inhomogeneous magnetic field, it is divided into two partial beams. Strictly speaking, the electric charge of the electrons interferes with this experiment, as it leads to an additional deviation of the trajectories. We shall simply ignore this in the following, since it does not matter here. With electrically neutral spin 1/2 particles – such as silver atoms – the experiment also works perfectly in reality (Fig. 6.6).

The deflection in the magnetic field in one direction or the other is always the same for the electrons. We usually simply say that the spin is $+1/2$ (spin $\uparrow$) or $-1/2$ (spin $\downarrow$), depending on the direction of deflection.

Whether a single electron is deflected in one direction or the other cannot be predicted – it happens randomly with certain probabilities. These probabilities can change if we tilt the magnetic field. Electrons that are deflected upwards or downwards in a vertical magnetic field with a 50% probability can for example be deflected 100% to the right in a horizontal magnetic field, if the spin state dictates this.

If we know the quantum mechanical spin state of an electron, we can calculate all these deflection probabilities in the different spatial directions and, of course, we can also simulate them on a computer. Everything seems clear here – at least that's the way it seems at first glance. But we should take a closer look, just as Feynman does in his lectures.

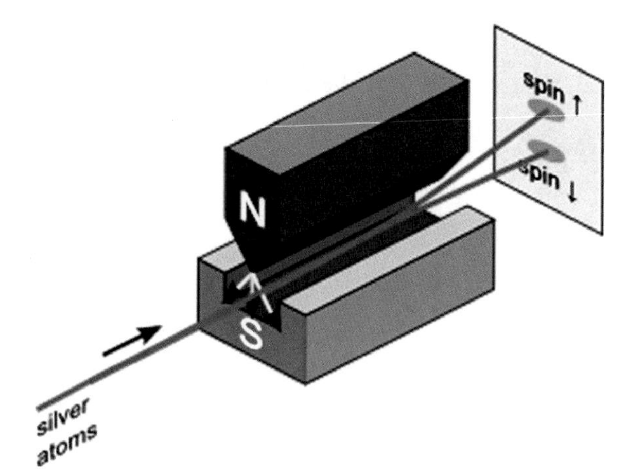

Fig. 6.6 In 1922, the German physicists Otto Stern and Walther Gerlach observed the deflection of silver atoms in an inhomogeneous magnetic field. In this famous Stern–Gerlach experiment, they discovered that the silver atoms are always deflected either upwards or downwards, depending on the spin orientation, but never in-between or laterally

The following considerations go into somewhat more detail and may require a little effort. Nevertheless, I would like to make them available to the interested reader, because they show that the strange dual nature of particles and waves is not the only strange feature of quantum mechanics. The theory puts our view of reality to an even more subtle test.

Entangled Spins and Spooky Action at a Distance

There is a phenomenon in quantum mechanics that does not exist in classical physics: so-called particle *entanglement*. This means that the behavior of one particle can directly influence the behavior of another particle and vice versa.

Let us look at a concrete example in which an electron source repeatedly emits two electrons in opposite directions. The spins of the electrons are in a special entangled state where the electrons behave like opposite twins: if one electron is deflected upwards in a vertical magnetic field, the other electron is deflected downwards, and vice versa (Fig. 6.7). Which of them is deflected upwards or downwards is not predetermined. The same behavior also applies to all other directions of the magnetic field, for example, the horizontal direction: if one electron is deflected to the right, the other electron will fly to the left. Furthermore, it does not matter how far apart the two electrons have already moved.

How does an electron know that its partner particle has been deflected upwards, for example, and that it must therefore orient itself downwards? Could the partner particle have sent it a message? Since the two electrons can theoretically even be light-years apart, but may have to decide simultaneously for opposite deflection directions, this is impossible: the message would have to propagate instantly over arbitrarily large distances. However, according to Einstein's special theory of relativity, the maximum allowed speed is the speed of light. This is why Einstein referred to this behavior of the two electrons as *spooky action at a distance* (spukhafte Fernwirkung).

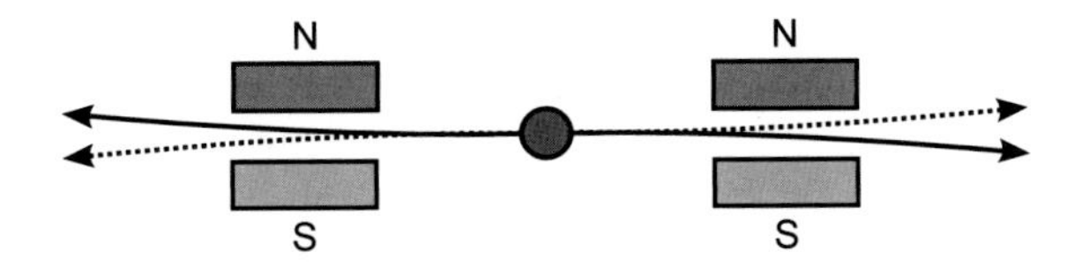

Fig. 6.7 Two electrons in a special entangled state are always deflected in opposite directions in a magnetic field

Local Hidden Variables: Is Quantum Mechanics Incomplete?

In a joint paper from 1935, Albert Einstein, Boris Podolsky, and Nathan Rosen regarded this spooky action at a distance as a proof that quantum mechanics must be incomplete. Since this long-distance action requires superluminal velocity and thus cannot exist according to the theory of relativity, each of the two electrons must already contain the information about how to react in the magnetic field. This information may be hidden from the outside, but it must exist if superluminal velocity is excluded and if physical systems behave locally – meaning that the behavior of one electron cannot depend on the simultaneous behavior of another electron that may be far away.

The requirement of local behavior was also required for our computer above. This seems so self-evident that it is easy to forget to mention it specifically. How could physics work if the outcome of a physical experiment on Earth could depend on what happens on Alpha Centauri at the same time? Einstein was therefore certain that nature must behave locally and that there must somehow be inner properties of electrons – also called hidden variables – which determine how they react in a magnetic field. Since these variables do not appear in quantum mechanics, quantum mechanics must be incomplete.

In order to work out what effect such an intrinsic property of electrons might have, Feynman considers things in the following way in his lecture. We imagine that one of the two electrons encounters an inhomogeneous magnetic field that is tilted to the left by some angle. We imagine further that an inner property of the electron determines whether it is deflected in or against the direction of the tilted field lines. Then we can capture the effect of this inner property in the following way.

We draw a circle and then draw a little white or black dot on it at the corresponding angle of rotation of the magnetic field, depending on the direction of deflection the electron chooses due to its internal property – white for one direction and black for the opposite direction of deflection. We can do this for different tilt angles of the magnetic field, because for all these angles the inner property determines the behavior of the electron. So the internal property of the electron is reflected in the color of the points we draw on the circle, which characterize the direction of deflection in a correspondingly tilted magnetic field. Opposite points must be colored in opposite colors, because if the electron in one magnetic field is deflected to the north pole, for example, it will be deflected to the south pole in the magnetic field rotated through 180°.

We now take the six tilt angles 0°, 60°, 120°, 180°, 240° and 300°, i.e., we tilt the magnetic field in 60° steps. The corresponding points on the circle then form a hexagon. How many possibilities do we have to color these six dots white or black?

Since opposite points must be colored in the opposite way, we can only freely choose the color for three of the six points, so that there are $2 \cdot 2 \cdot 2 = 8$ different possibilities – they are shown in Fig. 6.8. The circles above and below correspond to opposite behavior of the electrons. If an electron has an inner property that matches one of the upper circles, the partner electron must have an inner property that belongs to the circle below it, so that it always behaves in the opposite way.

We can now ask the following question. Suppose we randomly select one of the six magnetic field directions and measure the deflection of the electron. If we were to tilt the magnetic field another 60° to the left, what would be the probability that the same electron would also be deflected to the same magnetic pole in this magnetic field?

In our picture with the six dots on the circle, this question can be rephrased as follows: how often does it happen in a circle that two adjacent dots have the same color? The numbers in Fig. 6.8 give the answer: in most circles, four of the six adjacent pairs of points have the same color, and only in two circles are there no neighbors at all with the same color. So if we randomly select any point from any of the circles, the probability that the neighboring point has the same color counter-clockwise is a maximum of 4/6, which is 2/3.

For the electron, this means that *the probability that an electron would be deflected both times in the direction of the North Pole or both times in the direction of the South Pole with two magnetic fields tilted by 60° against each other, must be less than or equal to 2/3, due to the electrons local internal property which determines its deflection.*

In his talk *Simulating Physics with Computers*, Feynman presents this argument with photons instead of electrons, since it is experimentally easier to implement. Instead of using magnetic fields, we can work with a calcite crystal to measure the polarization of the photons, and the angle dependencies are also slightly different – but the overall result is the same.

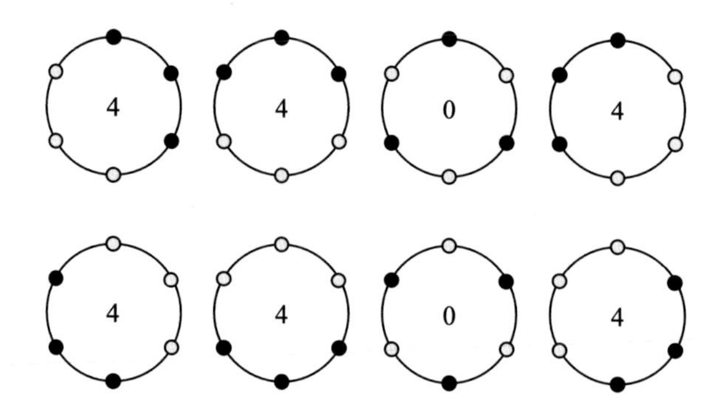

Fig. 6.8 Each circle corresponds to a possible inner property of the electron. The color of the individual points on the circles determines the direction in which an associated electron would be deflected in a correspondingly tilted magnetic field. The position of the point on the circle corresponds to the tilt angle of the magnetic field. The number in the center of the circle indicates the number of adjacent pairs of points of the same color

The Nonlocal Quantum World

The first arguments of this kind were presented by the Northern Irish physicist John Stewart Bell, who derived what is now known as *Bell's theorem* in 1964. Feynman manages not to mention John Stewart Bell at all in his lecture. But this was by no means due to a lack of respect for Bell's achievements, and Feynman did not want to adorn himself with foreign feathers either. Rather, Feynman focused on the idea itself, which he reconsiders in his own way and explains to his audience – therefore it was not always so important for him to mention the person who had first had the idea. This was quite typical of Feynman: he did not always give his colleagues the credit they actually deserved. With a colleague like Murray Gell-Mann, such negligence could sometimes lead to certain irritations, which is quite understandable – after all, scientists make a living out of their ideas and the reputation they give him.

So what does quantum mechanics say about our above argument? It is straightforward to calculate the corresponding probability within the framework of quantum theory. The result is that, *according to quantum mechanics, the probability that an electron would be deflected both times in the direction of the North Pole or both times in the direction of the South Pole with two magnetic fields tilted by 60° to one another is 3/4.*

This is in clear contradiction with the result we got above by considering hypothetical internal properties of the electron, since that argument said that this probability should be less than or equal to 2/3. The quantum mechanical prediction of 3/4 is higher by 1/12.

So which is right now? Is quantum mechanics right, or should we prefer our argument based on the circles with the black and white dots? Only experiment can answer this question. However, it is pointless to send the same electron successively through two mutually twisted magnetic fields, because the passage through the first magnetic field could change inner properties such as the orientation of its magnetic moment. The passage through the second magnetic field would then reveal nothing about its original properties.

This is where the partner particle comes into play: since it is guaranteed to behave in the opposite way, we can measure in a single experiment how the electron must react in two different magnetic fields. To do this, we send the electron through the original magnetic field and its partner particle through the magnetic field tilted by 60°. For example, if the partner particle is deflected towards the South Pole, we know that the electron would have been deflected towards the North Pole in the tilted magnetic field.

So what is the result of this experiment? It confirms precisely the prediction of quantum mechanics: with a large number of electron pairs, the electron and its partner are deflected in opposite ways in three quarters of all cases in the two magnetic fields tilted by 60°: one to the South Pole and the other to the North Pole or vice versa.

Conversely, it follows that the electron cannot have the local internal property we originally surmised. There can be no local hidden variable inside the electron which determines its behavior in each magnetic field. The correlated behavior of the two electrons cannot be explained in this way.

If we wish to continue to exclude superluminal velocities, it follows that nature does not behave locally. The behavior of two particles can be coupled to each other over arbitrarily long distances without them notifying each other and without already coordinating their behavior in the particle source and the particles then carrying this information with them. The outcome of a physical experiment on Earth can actually depend on what happens on Alpha Centauri at the same time. Einstein would certainly not have liked this result!

Shouldn't the whole of physics then collapse? How can we still make predictions when everything is interrelated? Fortunately, the situation is not that bad. After all, the two electrons must have been close to each other when they were emitted in the particle source in order to ensure their entangled spin state. For an observer who does not know this and sees one of the electrons arriving far away, the second electron plays no role: the observers sees only one electron that is sometimes deflected towards the North Pole and sometimes towards the South Pole. Only when another observer tells him about the behavior of the second electron (and this message can only have been transmitted at the speed of light), can he recognize the correlation between the two electrons. Therefore, it is not possible to transmit messages at superluminal velocity using the correlated behavior of the two electrons.

In principle, however, quantum mechanics allows nonlocal relationships between far away particles. For our classical local probabilistic computer this means that it cannot simulate the random behavior of quantum mechanics, because this is a nonlocal phenomenon.

A possible solution would be a computer that is itself based on the rules of quantum mechanics – a quantum computer or quantum simulator. Such quantum computers are currently under intensive investigation, because they are obviously capable of doing things that a normal computer cannot do. Feynman was also very interested in these capabilities, but in his days quantum computers were still largely a dream of the future. In his lecture, Feynman was also less concerned with the specific construction of such a computer, and more with the question of what can be learned about physics and quantum mechanics in particular, with the help of computers.

Feynman says at the end of his talk that he liked to amuse himself by squeezing the difficulties of quantum mechanics into a tighter and tighter corner, to get himself more and more worried about this particular feature. In this context, it seemed almost ridiculous that the whole problem could be squeezed down to a numerical question that one thing is bigger than another.

The randomness in quantum mechanics is therefore of a different kind than the randomness in local classical physics. But what is really behind it? "A very interesting question is the origin of the probabilities in quantum mechanics", writes Feynman, speculating that computer science could invent a different point of view to describe this. This hope has already been fulfilled to some extent, because research in the field of quantum computers is now shedding more and more light on the question of how quantum mechanics merges into classical physics and what actually happens during a measurement process. However, the outstanding issues have not yet been resolved. In any case, there is one point on which we can agree with Feynman: "Nature isn't classical, dammit, and if you want to make a simulation of nature, you'd better make it quantum mechanical, and by golly it's a wonderful problem, because it doesn't look so easy."

6.3 The Last Years and the Challenger Disaster

The last years of Feynman's life were overshadowed by his cancer. In 1978 – Feynman was sixty years old at the time – he was first diagnosed with cancer. The malignant tumor already weighed several kilograms and had destroyed a kidney and his spleen, so it had to be removed by major surgery. The operation was successful, but Feynman was ill for the rest of his life. He also knew that the likelihood of the cancer coming back was quite high – he probably didn't have more than ten years to live.

In the autumn of 1981, the cancer returned and a second major operation was necessary, which ended with some complications. It took some time for Feynman to recover from this ordeal. In the mid-1980s he fell seriously ill again. He knew that he would probably die within the next few years.

However, this did not prevent him from carrying on a normal life with the family he loved so dearly, and to pursue his interests with all the energy he could muster. He even embarked on an adventure that made him one of the most popular physicists of all time, even outside science.

A Shock to the World: The Challenger Accident

Feynman was recovering from his last operation when the worst accident in American space history occurred. On January 28, 1986, a good minute after launch, the Space Shuttle Challenger broke apart on its tenth mission at an altitude of about 15 km (Fig. 6.9). All seven astronauts on board lost their lives, among them the teacher Christa McAuliffe, who was to teach children live from space.

Like millions of other people, Feynman watched the launch on TV and saw the Space Shuttle explode in the sky. Until then he had never been particularly interested in NASA's shuttle program – he had the impression that it was of little scientific importance in itself. So apart from the tragedy that the Challenger disaster claimed seven lives, Feynman did not think too much about the accident.

However, a few days after the accident, Feynman received a call from the head of NASA, William Graham. Graham told Feynman that he had been a student at Caltech many years ago and had followed Feynman's lectures there. So he knew what an exceptional character Feynman was and how eager he was to solve problems. Graham now had the task of setting up a commission to investigate the Challenger accident, and Feynman seemed to him to be just right for the job.

When Feynman heard the investigation was to take place in Washington, his first reaction was: "How am I gonna get out of this?" He didn't want to go anywhere near Washington or having anything to do with government. But his friends said it was very important for the nation to find out the cause of the accident, and he should do it.

Fig. 6.9 Space Shuttle Challenger explodes shortly after take-off *Credit* NASA, Kennedy Space Center

His wife Gweneth also tried to convince him[2]: "If you don't do it, there will be twelve people, all in a group, going around from place to place together. But if you join the commission, there will be eleven people – all in a group, going around from place to place together – while the twelfth one runs around all over the place, checking all kinds of unusual things. There probably won't be anything, but if there is, you'll find it. There isn't anyone else who can do that like you can."

Finally, Feynman agreed and called Graham. He would put all his energy into the investigation for up to six months, but then he would leave the Commission no matter what the outcome. The next day Graham called back and told Feynman that he had been accepted as a Member of the Commission. It would be headed by former Foreign Minister William P. Rogers. The first meeting was scheduled to take place two days later in Washington.

Great, Feynman thought – then he had one more day to get at least a brief introduction to the technology of the Space Shuttle from his technically experienced friends and colleagues at the nearby Jet Propulsion Laboratory (JPL). This turned into a very thorough, high-speed, and intense briefing and Feynman learned a lot in a short time. He also got first hints about possible weak points of the Space Shuttle. For example, the second line of Feynman's meeting notes already contains information that the so-called O-rings showed scorching. These O-rings were large annular rubber seals used in the two solid rocket boosters. The rockets provided the necessary boost during the first two minutes of flight and were then jetti-

[2]See *What Do You Care What Other People Think.*

soned. They were composed of several cylindrical segments. At the joints, the O-rings were to ensure that no hot combustion gases could escape from inside the rockets. Scorch marks were not a good sign. Could it be that the O-rings did not always completely seal the joints? However, it was still too early to draw any more detailed conclusions. The Space Shuttle was technically highly complex and there were many other possible weak points.

Feynman on the Commission

After his lesson at JPL, Feynman took the night flight from California to Washington and arrived there at seven o'clock in the morning. The meeting took place in William Roger's office, where Feynman met the other members of the Commission. The only one of them Feynman already knew was Neil Armstrong, the first man on the Moon. The young physicist and astronaut Sally Ride was also among the commissioners, but Feynman only later realized that she was the first American woman in space – she had already flown the Space Shuttle Challenger twice, in 1983 and 1984. There was also a very handsome-looking gentleman in a uniform who was introduced to him as General Donald J. Kutyna (pronounced Koo-TEE-na). Feynman was no friend of formidable uniforms, but when at the end of the meeting the general asked about the way to the nearest metro station instead of getting into a limousine, Feynman revised his hasty judgment: "This guy, I'm gonna get along with him fine: he's dressed so fancy, but inside, he's straight." Feynman liked this general from the beginning, and would make no mistake about him. Kutyna also liked Feynman – he admired his intelligence, his honesty, his courage, and his tenacity, perfect for getting to the bottom of a problem. Such a person would be so useful on the Commission, he thought!

The first meeting was just an informal get-together. Feynman was somewhat disappointed, but at the same time he was relieved to learn that the investigation should be completed within 120 days. More meetings followed, but Feynman was not happy about the poor level of efficiency – he had learned more at JPL within a day. Feynman felt slowed down. Just sitting in meetings was a waste of time for him. He wanted to walk around and talk to the NASA engineers, which was not at all what Chairman William Rogers had in mind: "I was obviously quite a pain in the ass for Mr. Rogers", Feynman said. Later, he could better understand the complicated political situation in which Rogers had found himself, as chairman of the Commission – indeed, Rogers had a hard time keeping the investigation under control, in full view of the public and the press.

Graham finally made it possible for Feynman to talk to the NASA experts. From them he learned more details about the O-rings. When the enormous gas pressure builds up in the interior of the solid rocket boosters after ignition, the gaps between the individual segments expand within fractions of a second. The rubber of the O-rings must also expand fast enough to close the gap and maintain the seal. To do so, the rubber must be elastic enough.

If all the O-ring seals had leaked on all flights, the problem would have been immediately apparent. But only a few of the seals leaked on only some of the flights. According to Feynman, NASA therefore assessed these leaks as follows: "If one of the seals leaks a little and the flight is successful, the problem isn't so serious." It was like playing Russian roulette: you pull the trigger and the gun doesn't go off – so it must be safe to pull the trigger again.

A little later Feynman received the crucial hint from General Kutyna, who told him that he had been working on his carburetor that morning. Such a carburetor also contains O-rings as a seal, and they can leak in cold weather, as he knew from experience. Kutyna had the following idea: when the shuttle took off, the outside temperature was below freezing – it had been an extremely cold day. The coldest temperature at all other take-offs had been a good 10 °C. As a physicist, Feynman would certainly be able tell whether this low temperature would have any effect on the O-rings.

Feynman immediately realized what Kutyna was getting at: the cold would have made the rubber of the O-rings stiffen. This would mean that they might not have been able to react quickly enough to the expanding gaps at startup and seal them completely. It was such a good thing that Kutyna had given him this clue, Feynman thought, and continued: "A professor of theoretical physics always has to be told what to look for. He just uses his knowledge to explain the observations of the experimenters!"

Feynman thought that Kutyna had come up with this idea himself. What he didn't know was that Kutyna had received the hint from a befriended NASA astronaut and was looking for a way to bring the information into play without endangering the astronaut. It was by no means obvious that NASA would have welcomed someone "on the inside" spilling the beans. Such a person might soon have encountered career problems. It was thus important to protect the informant. And so Kutyna had set Feynman on the trail, because he was independent and would not allow NASA to dictate anything.

It was only in 2016 that Kutyna revealed who the informant had been: it was NASA astronaut Sally Ride, who was also a member of the commission. Sally had died in the meantime (in 2012), so that this revelation could no longer harm her. At the beginning of the Commission's work, she had given

Kutyna a NASA document describing a series of tests on the resiliency of the O-rings at different temperatures. It showed that they stiffened under cold conditions.

Feynman and Kutyna also found other signs of the failure of the O-rings. When they went to Graham, he showed them some interesting photos in which a flame could be seen growing from the right-hand solid rocket booster just a few seconds before the explosion. Other photos also appeared showing smoke coming out at the same place. Everything seemed to fit together.

Then an engineer of the company Morton Thiokol Inc. (MTI), where the sealing rings were manufactured, came to one of their meetings. He reported that his company's engineers had been very concerned for some time that low temperatures could lead to sealing problems. On the night before the launch, they had therefore recommended to NASA not to start below 10 °C. But under NASA's pressure, MTI had withdrawn the recommendation – against the advice of their own engineers. The Commission was shocked to hear this statement. Had NASA's management failed here, and knowingly ignored the warnings of the engineers?

O-Rings in Ice Water

Feynman wanted to get to the bottom of this. In the evening at the hotel he saw a glass of ice water in front of him and thought: "Damn it, I can find out about that rubber without having NASA send notes back and forth: I just have to try it! All I have to do is get a sample of the rubber." This would allow him in the public meeting about the sealing problems the next day to demonstrate what was really going on.

Feynman called Bill Graham, who was very cooperative as usual. He remembered that a model of the field joint with two samples of the rubber in it was to be shown at the meeting. Maybe a piece of rubber could be temporarily removed?

The next morning Feynman went to a hardware store, got the necessary tools, and went to Graham's office. There they actually managed to remove a piece of the sealing rubber from the model and see how the rubber would behave in ice water. The rubber was then put back into the model so that Graham could take it to the meeting. With his tools in his pocket, Feynman then went to the public meeting and sat next to General Kutyna.

But unlike in previous meetings, there was no ice water this time. The model was passed around, and Feynman's questions made it clear that if the

O-rings did not react within seconds to the opening of the joints during launch and reliably seal them, this would lead to a very dangerous situation. Unfortunately, he still needed the ice water to demonstrate in front of everyone how cold conditions could reduce the elasticity of the O-rings.

The model was passed on to General Kutyna and finally to Feynman. He took the tools out of his pocket and took the model apart to remove the O-ring piece. But the ice water was still missing. Then finally it came, and Feynman was ready to go. He squeezed the rubber in the C-clamp and put it into the ice water. After a few minutes it cooled down and Feynman made preparations to activate his microphone and speak out. General Kutyna suspected what Feynman was up to and whispered to him: "Co-pilot to pilot: not now!" Then he pointed to something in the briefing book that had previously been distributed and said: "When he comes to this slide, here, that's the right time to do it." Good thing General Kutyna, with all his political experience, had the right feeling for such situations!

Finally, the time had come and Feynman spoke up: "I took this rubber from the model and put it in a clamp in ice water for a while." Then he took the clamp out of the ice water and loosened it: "I discovered that when you undo the clamp, the rubber doesn't spring back. In other words, for more than a few seconds, there is no resilience in this particular material when it is at a temperature of 32° (Fahrenheit, i.e., 0 °C). I believe that has some significance for our problem."

The reactions were subdued at first, and the questions from the reporters during the lunch break showed that they did not understand what Feynman was getting at. Feynman later admits he was rather depressed that he wasn't able to make his point. But in the evening all the news channels showed his experiment, and the next day the newspapers explained everything perfectly. The journalists had meanwhile done their homework and realized what Feynman had been trying to explain to them. Now it fell upon them to explain it to the world. His unusual presentation had made Feynman a national hero overnight, bringing the truth to light.

Nature Cannot Be Fooled

Now it was no longer possible to hide the cause of the accident in a jumble of details, and people could ask why NASA had not recognized this, and had ignored the warnings of the engineers. Had NASA fooled itself into too much wishful thinking? Were there any other technical problems? Why hadn't management listened to its engineers? Was there too much political pressure not to postpone the planned launch? After all, a teacher was supposed to hold a lesson from space – a symbol of the nation's commitment to education, and a symbol of the high level of safety, hence also a very valuable public relations stunt for NASA. Feynman followed up these questions over the next few months and revealed a lot more, receiving many important hints from engineers and other NASA technical staff who were apparently not being heard by their own management.

Feynman was particularly upset about the risk assessment used by NASA's management. Indeed, it was claimed that the risk of a serious accident was only 1 in 100,000, which would mean that a shuttle could be launched every single day for three hundred years before a serious accident would be expected to occur. That was just ridiculous for such a complicated technical system. When Feynman interviewed the technical experts, they came to completely different results, which were usually between 1 in 50 and 1 in 200. "The management of NASA exaggerates the reliability of its product to the point of fantasy", wrote Feynman in the final report.

With hindsight, the final assessment seems quite realistic: between 1981 and 2011, a total of 135 shuttle flights took place and two serious accidents occurred: in 1986 the Space Shuttle Challenger was destroyed at launch, and in 2003 the Space Shuttle Columbia broke apart when it re-entered the Earth's atmosphere – at launch a piece of foam insulation broke off the external tank and struck the left wing of the orbiter, causing a hole in the heat shield. Two accidents on 135 flights – that was perhaps what should have been expected. Feynman found it all the worse that NASA, with its fantasy figure of 1 in 100,000, had led the public to believe a completely unrealistic safety assessment that had nothing to do with reality. The astronauts may well have been aware of the real risk. But would the young teacher Christa McAuliffe, who lost her life in the accident, also have entered the Challenger Space Shuttle if she had known about the real risk?

Feynman had some trouble getting his inconvenient findings – especially with regard to the lack of realism in NASA management – into the Commission's final report. Only when he threatened to remove his name from the report was he allowed to attach his results as an appendix. Feynman ends this appendix with the following words[3]:

> For a successful technology, reality must take precedence over public relations, for nature cannot be fooled.

---

[3]Feynman's appendix can be found, for example, at http://history.nasa.gov/rogersrep/v2appf.htm.

The main part of the Rogers Commission Report also clearly spelt out the critical facts, albeit more diplomatically than in Feynman's appendix. As a result, further shuttle flights were stopped and for two and a half years more than two thousand improvements were introduced to make the flights safer. The report, and certainly Feynman's clear statements, obviously had an effect. The solid rocket boosters never failed again, but in 2003 a hole in the fragile heat shield on the Columbia Space Shuttle led to the next accident. Unfortunately, the risk assessment by the technical experts proved to be correct, despite all efforts to ensure safety. In 2011, the last flight by Space Shuttle Atlantis ended the era of these rather successful, but nevertheless risky Space Shuttle flights.

This Dying Is Boring

When Feynman returned to Caltech after the end of the investigation in June 1986, he was quite exhausted. The strenuous work on the Commission had taken a lot of energy from him. In September, Feynman underwent another major operation.

In *Richard Feynman and The Connection Machine*, Danny Hillis recalls a conversation he had with Feynman during this difficult time when his health was getting increasingly worse, but he still had enough strength to roam the hills above Pasadena together with Hillis. They were exploring an unfamiliar trail and Feynman, who was recovering from his operation, had to walk slowly. In his typically witty way, he told how he had read everything about his illness and surprised his doctors by predicting their diagnoses and his own chances of survival. Hillis realized how bad Feynman's condition really was, and was quite concerned. "Hey, what's the matter?" Feynman asked. Hillis replied that he was very sad that Feynman was going to die. "Yes," Feynman sighed, "that bugs me sometimes too. But not so much as you think. When you get as old as I am, you start to realize that you've told most of the good stuff you know to other people anyway."

The next serious operation took place about one year after this operation, in October 1987. In November Feynman appeared for the last time in public – the progressing cancer was now increasingly taking its toll.

In December 1987 Feynman received a call from Jagdish Mehra, an Indian–American physicist and historian of science. Mehra and Feynman had known each other for many years, because Mehra was in the process of writing a scientific biography about Feynman – in 1994 it would appear under the title *The beat of a different drum – the life and science of Richard*

Feynman. Mehra had already learned a lot about Feynman's life at their previous talks, but he still had some unanswered questions. He therefore suggested visiting Feynman in Pasadena, but Feynman was not enthusiastic at first – he was too tired and depressed to talk about his past again. But four days later he called Mehra and changed his mind: he had read Mehra's book about Schrödinger and liked it. If Mehra still wanted to come, he would be welcome.

Mehra suggested the beginning of March, but Feynman hesitated: "I don't know, it may be too late." Mehra was alarmed and changed his plan. On January 9, 1988 he arrived in Pasadena and visited Feynman the next day. Feynman's emaciated appearance shocked him – it was clear that Feynman was not well and was suffering from severe pain. Nevertheless, they agreed to meet in the mornings for a while, where Mehra could ask questions and record the answers on tape. At lunch and in the afternoon Feynman wanted to relax and asked Mehra to tell him stories in return – as we know, Feynman loved good stories.

This gave Mehra a unique overview of Feynman's life and the topics that inspired and moved him. Feynman emphasized how much he loved physics – it was his hobby, his profession, and his pleasure at the same time. One of the things he was particularly proud of was his discovery of how the breaking of mirror symmetry, e.g., in beta decay, can be described theoretically. This was the only time in his life that he had discovered a new law of nature. A special sense of achievement was also the invention of the path integrals, which eventually led him to his formulation of QED. He would also have liked to solve the problem of superconductivity, but in the end he had not succeeded in doing so.

Asked about his scientific role models, Feynman named Sadi Carnot, James Clerk Maxwell, and especially Paul Dirac. We have already learned a lot about Dirac in this book. He was Feynman's outstanding hero and Feynman would have liked to make a discovery as significant as the Dirac equation. The situation was very similar with Maxwell, whose equations had put electrodynamics on a solid theoretical foundation and also combined it with optics.

The French physicist and engineer Sadi Carnot is less well known. He was the founder of thermodynamics in 1824 with his theoretical description of an idealized steam engine. Tragically, Carnot died of cholera in 1832, at the age of just 36. In Vol. I, Chap. 44 of his *Feynman Lectures*, Feynman deals several times with Carnot's pioneering ideas, which eventually led to the second law of thermodynamics. He admired Carnot for his achievement and writes: "In fact, the science of thermodynamics began with an analysis, by the great engineer Sadi Carnot, of the problem of how to build the best and most efficient engine, and this constitutes one of the few famous cases in which engineering has contributed fundamentally to physical theory."

Feynman also said that he had learned a lot from other people. In his childhood and youth, these were of course his father and his school teacher Abram Bader. Later, at MIT, he was influenced by people like Phillip Morse and John Slater, and in Princeton by his mentor, the creative eccentric John Wheeler. Einstein and Pauli were remembered by Feynman at least as nice gentlemen, even though his contact with them had been rather short. And we must not forget Hans Bethe. He had met him in Los Alamos and followed him to Cornell University.

At Caltech, with its focus on the natural sciences and engineering, Feynman felt particularly at home. There he had made many friends. Murray Gell-Mann was a wonderful and very inspiring colleague – a remarkable statement, considering their relationship, which was often dominated by rivalry. Despite all the differences in personality, both Gell-Mann and Feynman obviously knew exactly how outstanding the other was.

Feynman emphasized his good fortune in meeting such a nice woman as Gweneth, who had given him such a loving home, and the children he was so proud of. He then told of a plan that he and his friend Ralph Leighton had been pursuing for many years, but which he could no longer accomplish due to his poor health: to visit *Tannu Tuva*. Feynman had known Ralph Leighton since Leighton had been a child. Ralph was more than thirty years his younger. He was the son of Caltech physicist Robert B. Leighton, with whom Feynman was also friends. Robert Leighton had played a decisive role in the publication of the Feynman Lectures.

With Ralph Leighton, Feynman shared the passion for drumming on his beloved bongos. Moreover, both enjoyed adventures and faraway countries. One evening in the summer of 1977, Feynman had asked his friend if he had ever heard of Tannu Tuva – Feynman knew this place from various exotic stamps he had collected. Leighton had no idea, so they searched for information. They found the small remote country in southern Russia near Mongolia, deep in the heart of Asia. At that time Tannu Tuva was part of the Soviet Union, and known as Tuvan ASSR, but today it is an independent republic and member of the Russian Federation. The region was particularly important for the Soviet Union because of its uranium deposits.

When Feynman and Leighton saw that Tuva's capital city was called *Kyzyl*, their decision was clear: a city without a single vowel in its name must be interesting. Apparently, they did not consider the two "y"s in Kyzyl as vowels. They should have, because they are indeed the English transcription of a Russian vowel. Murray Gell-Mann would certainly have known this and would have enjoyed pointing out this error to Feynman.

In any case, Feynman and Leighton were determined to visit the city with the strange name. They pursued their plan for over ten years without ever actually overcoming all the obstacles. On the one hand, this was a sparsely populated region right in the middle of Central Asia and very isolated, so there was no way of simply taking the next plane. On the other hand, the Cold War was raging, and Feynman had worked as an American on the construction of the first nuclear bomb – so the Soviet Union had little interest in letting someone like that travel near its large uranium deposits.

Leighton, who also published the books *Surely You're Joking, Mr. Feynman!* and *What Do You Care What Other People Think?* based on interviews with Feynman, describes their long-standing efforts to finally reach Tannu Tuva in his book *Tuva or Bust! Richard Feynman's Last Journey*. The story wonderfully reflects Feynman's zest for life, his curiosity, and the fun he had in solving problems. He and Leighton struggled to gather information about the little known country – the Internet did not yet exist. They made contacts there and even wrote a letter in the Tuvan language. They also managed to bring an exhibition of art objects from Tuva to the USA. When it finally looked as if the trip would be possible, after many years of effort, Feynman was too ill to start the journey.

Mehra left Feynman on January 27 and knew that he would never see this extraordinary man again. A few days later, Feynman's still functioning kidney failed as the cancer continued to progress. Everyone knew it was coming to an end. Richard told his sister Joan and his wife Gweneth that he would like to avoid dialysis because this would only prolong his suffering for nothing. The two women came to him at the clinic and agreed. Soon Feynman fell into a coma, from which he awoke briefly and uttered his last words: "This dying is boring."

Feynman died on the late evening of February 15, 1988 at the age of 69. Gweneth, who had also had cancer for a long time, followed him on 31 December 1989 – she was only 55 years old. They were buried together at Mountain View Cemetery in Altadena, California. Only a simple plaque marks their grave. On it you will find the text:

IN LOVING MEMORY
FEYNMAN
1918 RICHARD P. 1988
1934 GWENETH M. 1989

6.4 Feynman's Legacy

What remains of Richard Feynman, the man who had such a decisive influence on physics in the mid-twentieth century? What were his greatest achievements, and how will he be remembered by posterity?

Most physicists know him for the Feynman diagrams, a tool that pervades much of modern physics. Just how widespread these diagrams were, even in Feynman's own lifetime, is illustrated by the following anecdote recounted by the author Al Seckel.[4] Feynman had a van with his diagrams painted all over it. On one of his trips, he was standing with his van in front of a snack bar somewhere in the Midwest when a young man came up to him and asked why he had Feynman diagrams all over his van. "Because I AM Feynman," was his answer. The young man went "Ahhhhh….."

Feynman himself was not quite so impressed by his diagrams and his other work on QED, for which he received the Nobel Prize. He had the feeling that he had not discovered anything fundamentally new, but only developed a new calculation method, although it was admittedly very practical.

But he was particularly proud of his less well-known work on the breaking of mirror symmetry in weak decays. Feynman had had the feeling that, for the first time in his life, he had discovered a new law of nature that nobody else knew. Only later did he find out that Gell-Mann, Sudarshan, and Marshak had also found this natural law by other means. But he didn't care – it was not so important for him to be the first.

Feynman has also left his mark in completely different areas of physics, for example in the quantum description of superfluid helium. His ability to understand the essential characteristics of a physical system intuitively was once again of great benefit to him. However, we should not forget that this intuition, which sometimes seemed like magic, was usually the result of wide-ranging investigation and detailed calculations that Feynman had made beforehand. He kept scribbling, calculating, trying out various ideas, rejecting them, and working his way step by step to the heart of each problem.

[4]See http://www.fotuva.org/online/frameload.htm?/online/seckel.htm.

Feynman had proceeded in the same way with the problem of quantum gravity, which he returned to over and over again for many years. He had studied the subject intensely, experimenting with his ideas on massless spin-2 gravitons and always trying to push further forward. His colleagues were often surprised by Feynman's knowledge on this subject and the breathtaking speed with which he was able to ignite a firework of new ideas. They simply did not realize that he had been working intensely on the topic for years.

Hardly any other physicist had been able to play a leading role in current research for several decades as Feynman had. He was already fifty years old when his parton model succeeded in explaining the deep-inelastic electron–proton collisions at SLAC and thus helped to establish the quark model of his colleague Gell-Mann.

It's interesting to note what Gell-Mann himself thought was Feynman's greatest achievement. In *Dick Feynman – The Guy in the Office down the Hall* (*Physics Today*, Feb. 1989), he writes after Feynman's death that the path integral formulation of quantum mechanics is probably a truly fundamental advance in physical theory. Feynman had already developed the path integrals in his doctoral thesis, following on from an idea of Dirac to clarify the role of the action in quantum mechanics. At that time Feynman regarded these path integrals merely as a reformulation of the existing quantum theory which was particularly useful in the context of relativity theory, but nevertheless contained nothing really new.

In contrast, Gell-Mann argues that the path integral formulation of quantum theory is probably more fundamental than the usual standard formulation. This becomes apparent if we try to describe the entire universe as a quantum system, including gravity. In addition to the paths of all particles, you need to take into account all conceivable space-time curvatures that the universe can pass through in principle. This means that the world can no longer be described as a quantum system existing in classical space and time, as is done in conventional quantum theory. Space and time themselves become a quantum system. The path integral can handle something like this, since it simply assigns a quantum amplitude to every conceivable space-time geometry – in other words, to every possible history of the universe, no matter how curved and complex the geometry may look.

Feynman's path integral has indeed become a standard tool in many areas of theoretical physics. Perhaps one day the outcome of his doctoral thesis will turn out to be the most fundamental scientific work that Feynman ever did in physics. The path integral could be the real foundation of quantum theory and thus of physics as a whole. Perhaps Feynman's path integrals are even the only truly fundamental way to formulate a quantum theory.

Feynman came up with his ingenious ideas because he loved to tackle every problem, whether significant or insignificant, in a completely new way. He would always want to play with the problem and reconsider it in his own way. Gell-Mann, who was Feynman's office neighbor at Caltech for about thirty years, put it basically like this: "Richard's approach went along with his extraordinary efforts to be different. Of course any of us engaged in creative work has to shake up the usual patterns in some way in order to get out of the rut of conventional thinking, and find a new and better way to formulate some problem. But with Dick, *turning things around* and being different became a passion. The result was that on certain occasions, he could come up with a remarkably useful innovation. But on many other occasions when the usual way of doing business had its virtues, he was not the ideal person to consult."

This trait made Feynman an outstanding visionary in many different areas, including nanotechnology and computer science. When Feynman was asked to take on the apparently standard task of giving the introductory lectures for freshmen at Caltech, it was clear that this would not be one of the usual standard lectures. Feynman's qualities as a storyteller and entertainer mixed with his intention to open up a new and profound approach to physics that went well beyond what is usually offered. In this way, his famous Feynman Lectures emerged, and their popularity remains intact to this day.

Feynman himself was at first rather sceptical about the success of his lectures. In his talk with Mehra shortly before his death he says[5]: "At the end of the two years (1961–1963) I felt that I had wasted two years, that I had done no research during this entire period and I was muttering to this effect." But then Robert Walker, who had given the lecture on mathematical methods, approached him and said: "Someday you will realize that what you did for physics in those two years is far more important than any research

[5]See http://www.feynmanlectures.info/popular_misconceptions_about_FLP.html.

you could have done during the same period." At the time Feynman had said he was crazy, but that had changed in the meantime. Since the books still enjoyed great popularity even after many years, he could no longer deny that they really were a major contribution to the world of physics. He even came to believe that creating the Feynman Lectures was one of the best things he ever did.

Even outside the physics community, Feynman was able to convey his enthusiasm for physics to his listeners and readers in many lectures and books. His popular book *QED – The Strange Theory of Light and Matter* is an impressive example of how a complicated theory can be explained to a broad audience without the use of formulas and without misleading simplifications.

Many people remember Feynman, not for his physics but rather for the countless anecdotes he became embroiled in. Ralph Leighton in particular has tried to collect these stories and preserve them for posterity – the resulting books *Surely You're Joking, Mr. Feynman!* and *What Do You Care What Other People Think?* have become bestsellers. Murray Gell-Mann was less enthusiastic about this: "He (Feynman) surrounded himself with a cloud of myth, and he spent a great deal of time and energy generating anecdotes about himself."

Of course, Feynman never hesitated to put himself forward, but this was far less about his personal fame, than fascinating people and generally entertaining them with good stories. Feynman had always loved to tell stories since his childhood, and he had a good sense of suspense, and how to deliver a punch line. In addition, many other people who knew Feynman had interesting experiences with him and told the corresponding anecdotes themselves. Feynman was just a very inspiring and unusual personality – the stories came about almost of their own accord.

What Gell-Mann appreciated most about Feynman was his direct style, completely lacking in pomposity. Feynman hated it when his colleagues dressed up their sometimes rather modest contributions in fancy mathematical formalism and surrounded them with all sorts of pomp. When he had this impression in a seminar, he could become quite unpleasant and sometimes offended some of his colleagues – in a way that could not always be justified. Feynman himself made every effort to present his own ideas, which were often powerful and refined, as straightforwardly and simply as possible. He wanted people to understand what he had to say.

Feynman's outstanding characteristics were his honesty, his frankness, and his scientific integrity. Every idea, no matter how beautiful and convincing, had to undergo a merciless verification by experiment. There could be no whitewashing: if there were clear differences between theory and experiment, the theoretical idea had to be abandoned. "*The first principle is that you must not fool yourself – and you are the easiest person to fool*", is one of Feynman's most famous quotes. If Feynman himself had made a mistake, he had no problem admitting it. The truth had absolute priority over personal sensitivities.

It was above all this characteristic that made him a well-known and admired person by the end of his life. It was intolerable for Feynman to see NASA management turn a blind eye to the existing and well-known risks of shuttle flights and ignore the warnings of its own experts. The truth had to be revealed, and Feynman found a way to do so with his experiment in front of running cameras. Not many people would have read the Commission's final report, but everyone could immediately understand the pictures of Feynman squeezing the rubber piece of the O-ring in the C-clamp, putting it in ice water and then demonstrating how it lost its elasticity in the cold. Feynman thus made the point very clear and nobody could ignore it anymore.

How satisfied was Feynman himself with his life's work? Could he have done more if he had read his colleagues' work more often, instead of always having to rediscover everything himself? Maybe so, but that was not his style. Great researchers must be wilful to a certain extent, otherwise they will remain trapped by the ideas of others, on paths that have already been trodden. Feynman knew like no other how to break away from these paths and go his own way – with all the advantages and disadvantages that this entailed.

In *The Pleasure of Finding Things Out*, Feynman mentions that people often asked him if he was looking for the ultimate laws of physics – what we often call the world formula. But that wasn't his goal! It was enough for him simply to find out more about the world, and if there really was a simple ultimate law of nature that explained everything, so be it – that would be very nice to discover. Nature is there and she's going to come out the way she is. Feynman has done much to lift the veil behind which the true essence of nature is hidden. But we do not yet know that much, and what we know we usually do not know with certainty. Feynman preferred to live with this uncertainty rather than clinging to any beliefs that might later turn out to be untenable. For him, doubt was a central part of science, and he was not afraid of the uncertainty that this entails. In *The Pleasure of Finding Things Out* he puts it this way:

> I don't feel frightened by not knowing things, by being lost in a mysterious universe without having any purpose, which is the way it really is so far as I can tell. It doesn't frighten me.

Glossary

Action If the Lagrangian (kinetic minus potential energy) of a system is integrated over a period of time, the so-called action for the associated motion is obtained. The actual motion can be determined by the principle of least action.

Advanced waves In classical electrodynamics, these are electromagnetic waves that formally propagate backwards in time from a vibrating charge. Viewed in the normal time direction, these waves approach the charge and are absorbed by it.

Amplitude Current height of a wave in one place. Positive amplitudes correspond to wave crests, negative amplitudes correspond to wave troughs. The amplitude of a quantum wave is not exactly described by a single value, but by a so-called complex number, which can be illustrated as an abstract arrow or clock hand having length and angle of rotation.

Angular momentum Indicates the quantity of rotational motion of a spinning object, just as the momentum indicates the quantity of motion of a linearly moving object.

Antimatter Antimatter consists of antiparticles. Thus, the core of an antiatom is made of antiprotons and antineutrons and the shell is composed of positrons. When matter and antimatter meet, they annihilate each other and dissipate, transforming all their mass into energy according to the formula $E = mc^2$.

Antiparticles Antiparticles have (with a few very small exceptions) exactly the same properties as the associated particles, but opposite charges. With regard to the weak interaction, they also behave in a mirror-inverted way to each other. A particle and its antiparticle can be formed together from energy (e.g., in the form of photons) and can also annihilate each other again.

Atomic nucleus Atomic nuclei form the tiny massive centers of atoms. They consist of protons and neutrons, which are held together by the strong interaction. Almost the entire mass of an atom is located in its atomic nucleus.

J. Resag, *Feynman and His Physics*, Springer Biographies,
https://doi.org/10.1007/978-3-319-96836-0

Baryons Baryons are strongly interacting particles that consist of three quarks. Protons and neutrons are the most common baryons, but there are many others, all of which are unstable and decay after a very short time.

Beta decay The decay of a neutron into a proton plus electron and an electron-antineutrino is called beta decay. This decay is caused by the weak interaction and can occur both with free neutrons and with neutrons within a radioactive atomic nucleus. In certain atomic nuclei, the decay of a proton into a neutron plus positron and an electron-neutrino is also possible, and this is then called beta-plus decay.

Bose–Einstein condensation At very low temperatures, bosons (particles with integer spin, e.g., helium-4 atoms) tend to collectively occupy the lowest quantum state. If this happens, we speak of Bose–Einstein condensation.

Bosons Particles with integer spin, for example photons and gluons (spin 1), gravitons (spin 2), or helium 4 atoms (spin 0).

Color charge The color charge determines how strongly the strong interaction affects a particle. Only quarks and gluons carry a color charge. The possible values of color charges are "red", "green", and "blue" or "antired", "antigreen", and "antiblue". These values have nothing to do with optical colors.

Confinement Quarks and gluons are subject to confinement, i.e., they can only exist inside baryons and mesons. Free quarks and gluons do not exist.

CP violation If we look at a fundamental physical process in a mirror (P for parity) and replace all particles by antiparticles (C for charge conjugation), we almost always see a physically possible process which takes place in the same way. There are a few processes where this is not exactly the case. This is referred to as CP violation. CP violation is partly responsible for the fact that a little more matter than antimatter was created in the Big Bang.

CPT theorem If we look at a fundamental physical process in the mirror (P for parity), replace all particles by antiparticles (C for charge conjugation), and then let time run backwards (T for time reversal), we always see a physically possible process. This fundamental CPT theorem is a general consequence of relativistic quantum theory.

Dirac equation The Dirac equation is a relativistic differential equation for the quantum waves of particles with spin 1/2, for example electrons. Particles with spin 0, on the other hand, are described by the relativistic Klein–Gordon equation.

Electric charge The electric charge determines which force acts on an object in an electromagnetic field. It is a conserved quantity, so it can neither disappear nor arise from nothing.

Electromagnetic interaction The electromagnetic interaction describes the forces that electrically charged objects exert on each other due to their charges. This also includes magnetic forces and all optical phenomena, since light is an electromagnetic wave.

Electron According to our current understanding, the electron is an elementary, stable, point-like particle with a negative elementary charge and spin 1/2. Electrons form the shells of atoms, being about two thousand times lighter than the protons and neutrons in the atomic nucleus.

Elementary charge The electric charge of free particles is always an integer multiple of the elementary charge. For example, the electron carries a negative and the proton a positive elementary charge.

Energy Energy is a quantity that is conserved in sum in all processes in nature. According to Noether's theorem, this is a consequence of the fact that, in the fundamental laws of nature, all instants of time are equivalent (time invariance). In contrast to momentum, energy is a scalar quantity, i.e., a number and not a vector. According to the special theory of relativity, masses must also be included in the energy balance, because mass is a form of energy.

Fermions Particles with half-integer spin, such as electrons, neutrinos, and quarks. Fermions are subject to the Pauli exclusion principle and therefore try to avoid each other.

Feynman diagram Each Feynman diagram represents one way a physical quantum process can proceed. For each Feynman diagram, a formula can be provided to calculate the contribution with which this option contributes to the overall process. Feynman first developed this vivid calculational tool for QED.

Flavor The six types of quarks are distinguished by their flavor. The possible flavors are up (u), down (d), strange (s), charm (c), bottom (b), and top (t).

Force A force is an external influence that leads to an acceleration, or more precisely to a change in the momentum of a moving object.

Frequency For a wave, the frequency tells us how fast the wave oscillates up and down at a certain point. In quantum mechanics, the frequency of a wave is proportional to the energy of the associated particle.

***g*-factor** This indicates the strength of the magnetic moment of a particle by comparing it with the magnetic moment generated by a classical circular current with the same angular momentum.

General theory of relativity Albert Einstein's theory of gravity, which extends the special theory of relativity. Gravitational effects propagate at a maximum speed equal to the speed of light. As a result, gravity is no longer described by forces, but by the curvature of space and time.

Gluon Neutral massless particle with spin 1, which mediates the strong interaction – analogous to photons for the electromagnetic interaction. Unlike photons, which can also exist as free particles, gluons are confined inside mesons and baryons, just like quarks (confinement).

Gravitation Gravitation or gravity is an attractive force between objects solely due to their mass or energy.

Gravitational constant The gravitational constant G describes how strongly two masses of one kilogram attract each other when separated by a distance of one meter. The value is 6.674×10^{-11} newton, so the corresponding attraction is very small, in fact barely perceptible. Thus gravity is much weaker than the other three fundamental interactions (the electromagnetic, weak, and strong forces).

Graviton A hypothetical, electrically and color-neutral, massless particle with spin 2, which mediates mass attraction in a quantized gravitational theory. Gravitational waves consist of a large number of individual gravitons. Feynman used gravitons as the starting point for his theory of quantum gravity.

Interaction There are four fundamental interactions between particles in nature: the weak, strong, and electromagnetic interactions and gravity.

Kaons, K mesons The four kaons (one positive, one negative, and two electrically neutral) are about half as heavy as a proton. These mesons consist of an up or down quark or antiquark and a strange antiquark or quark. Therefore, they have a strangeness different from zero, and can only decay through the weak interaction, which explains their comparatively long lifetimes. Their decays into two or three pions played an important role in the discovery that the mirror symmetry of nature is violated in the weak interaction.

Klein–Gordon equation The Klein–Gordon equation is a relativistic differential equation for the quantum waves of particles with spin 0, whereas particles with spin 1/2, such as electrons, are described by the Dirac equation.

Lagrangian In classical mechanics, the Lagrangian of a system (e.g., a particle in a force field) is the difference between its kinetic and potential energy. If the Lagrangian is integrated over a period of time, this yields the so-called action for the associated motion.

Leptons The electron, muon, and tauon together with the three associated neutrinos form the group of leptons. Together with the six quarks and the interaction particles (photon, gluon, $W^{\pm}$ and Z boson), the six leptons form the fundamental constituents of matter and its interactions in the standard model of particle physics. Unlike quarks, leptons are not subject to the strong interaction.

Lorentz contraction The Lorentz contraction or relativistic length contraction means that, if an observer measures the length of a moving object, this length will appear shorter in the direction of motion than if the object were stationary. The Lorentz factor indicates the extent of this length contraction, which is predicted by the special theory of relativity.

Lorentz factor The Lorentz factor γ indicates how much the inertia of a moving object increases with increasing speed, according to the special theory of relativity. It also indicates the extent of time dilation and Lorentz contraction. At speeds well below the speed of light, γ is almost equal to one and can be neglected. The more the speed approaches the speed of light, the greater the value of γ.

Magnetic moment This indicates the strength of a magnet, such as a rod magnet or compass needle. Particles with spin other than zero, such as the electron, often have a magnetic moment, so they behave similarly to a compass needle.

Mass The mass of an object which is stationary or moving much more slowly than light indicates its inertia, i.e., its resistance to acceleration (inertial mass). According to the special theory of relativity, mass is nothing else than the rest energy ($E = mc^2$), except for the conversion factor c^2. In addition, the mass also determines how much gravity acts on an object (gravitational mass). According to Einstein's general theory of relativity, inert and heavy masses are indistinguishable.

Maxwell equations The Maxwell equations are the fundamental equations of the electromagnetic interaction. They describe the mutual influences between electric charges and currents, together with the electric and magnetic fields they generate, as well as all phenomena caused by electromagnetic waves.

Mesons Strongly interacting particles consisting of a quark and an antiquark. Pions, kaons, and rho mesons are the most common mesons. All mesons are unstable and decay within fractions of a second. They are formed, for example, by nuclear collisions in cosmic radiation and particle accelerators.

Mirror symmetry Mirror symmetry (parity conservation) means that you cannot decide whether you are viewing a fundamental physical process in a mirror or not – the mirror counterpart of any physical process is equally possible. In nature, mirror symmetry is broken in processes involving the weak interaction, e.g., in radioactive beta decay – a phenomenon that Feynman played a decisive role in explaining.

Momentum The momentum indicates the quantity of motion of a moving object. One can also imagine the momentum as a stored impact or impulse. Momentum is conserved in sum in all processes in nature. According to Noether's theorem, this is a consequence of the fact that all points in space are of equal importance in the fundamental laws of nature. In contrast to the energy, the momentum is a quantity that points in the direction of motion, i.e., it is represented by a vector.

Muon From today's point of view the muon is an elementary, point-like particle with negative charge and spin 1/2. It thus has similar properties to the electron, but is about 200 times heavier and therefore unstable. Muons can be formed by cosmic radiation in the upper atmosphere and decay in fractions of a second.

Neutrino Neutrinos are almost massless neutral particles with spin 1/2 which are only subject to the weak interaction and therefore hardly interact with matter at all. Thus, neutrinos can often pass through the entire Earth without being hindered. Neutrinos are formed in large numbers in nuclear reactions, for example during beta decay, but also in cosmic radiation or during nuclear fusion inside the Sun. There are three different types of neutrinos: electron, muon, and tauon neutrinos, each with its own antiparticle.

Neutron Electrically neutral baryon consisting of one up quark and two down quarks (udd). Protons and neutrons are the constituents of atomic nuclei. Free neutrons are not stable – they decay into a proton, an electron, and an electron-antineutrino via beta decay with a half-life of about 10 minutes. In atomic nuclei, however, neutrons are usually stable if their number does not become too great.

Noether's theorem Noether's theorem states that conserved quantities result from physical symmetries. For example, the conservation of energy is a consequence of the fact that in the fundamental laws of nature all instants of time are equivalent (time invariance).

Parity violation Violation of mirror symmetry (see Mirror symmetry).

Pauli exclusion principle This is the property that fermions (particles with half-integer spin) avoid each other in the sense of never occupying exactly the same quantum state. One consequence is the structure of the electron shells in atoms, as electrons have spin 1/2 and are therefore subject to this principle.

Photon Neutral massless particle with spin one mediating the electromagnetic interaction. Electromagnetic waves like those constituting visible light consist of a large number of individual photons, which are therefore also called light quanta.

Pions The three pions π^+, π^0, and π^- are the lightest mesons, with mass about 0.15 times the proton mass. They consist of up and down quark–antiquark pairs and decay within a fraction of a second.

Planck constant The Planck constant h or $\hbar = h/(2\pi)$ links particle properties (energy, momentum) with wave properties (frequency, wavelength) in quantum mechanics. The Planck constant is the fundamental physical constant of quantum theory.

Planck units (Planck length, Planck time, Planck energy, Planck mass) In a relativistic theory of quantum gravity, the gravitational constant G, the speed of light c, and the Planck constant h can be combined to form new quantities that can serve as natural reference quantities (Planck units) for times, lengths, masses, and energies. These quantities also indicate the scale at which a quantum theory of gravity is expected to become important.

Positron Antiparticle of the electron. Positrons have the same mass and spin as electrons, but the opposite charge.

Principle of least action If one considers for a fixed period of time not only the actual motion but also imagined motions of a physical system and calculates the action for each one, the actual motion will be the one with the smallest action. Conversely, the actual motion can be calculated from this property.

Probability amplitude The probability amplitude is the amplitude of a quantum wave. The squared absolute value of the probability amplitude represents the probability for a certain event, e.g., the probability of finding a particle at a given location.

Proton Positively charged baryon consisting of two up quarks and one down quark (uud). Protons and neutrons are the constituents of atomic nuclei. The proton is the only baryon that is also stable as a free particle and does not decay (at least according to today's knowledge).

Quantum chromodynamics (QCD) QCD is the quantum theory of the strong interaction. The fundamental interaction particles of QCD are the gluons that hold the quarks in the mesons and baryons together.

Quantum electrodynamics (QED) QED is the quantum theory of the electromagnetic interaction. The fundamental interaction particle of QED is the photon. Richard Feynman played a decisive role in the formulation of QED with the help of his Feynman diagrams.

Quantum field theory If the principles of special relativity theory are incorporated into quantum mechanics, we obtain a quantum field theory. The best-known quantum field theory is quantum electrodynamics (QED), in which Feynman played a decisive role.

Quantum gravity Quantum gravity is a relativistic quantum description of gravitation. The consistent formulation of such a theory is still the subject of much research. Feynman took the first steps in this direction, assuming the existence of gravitons as the interaction particles of gravity.

Quantum mechanics This describes the dynamics of particles using quantum waves.

Quantum wave In quantum mechanics, particles no longer move on well-defined paths, but as quantum waves. The more intense a quantum wave at a given location, the more likely we are to find the particle there.

Quarks From today's point of view, quarks are elementary point-like particles with a fractional charge ($\pm 1/3$ or $\pm 2/3$) and spin ½. They are the constituents of baryons and mesons. There are six different types of quark (flavors), which are called up (u), down (d), strange (s), charm (c), bottom (b), and top (t). For each quark there is an antiquark with the opposite value of electric and color charge. Unlike leptons, quarks are subject to the strong interaction that binds them together to form baryons and mesons. Free quarks do not exist (confinement).

Retarded waves In classical electrodynamics, these are electromagnetic waves that propagate forward in time from a vibrating charge, analogous to the electromagnetic waves emitted by an antenna.

Rho mesons (ρ mesons) The three rho mesons ρ^+, ρ^0, and ρ^- are relatively heavy mesons with masses of about 0.83 times the proton mass. Like the lighter pions, they consist of up and down quark antiquark pairs and decay extremely fast via the strong interaction – much faster than pions or kaons.

Schrödinger equation The Schrödinger equation is the fundamental dynamical equation of non-relativistic quantum mechanics. It describes how a quantum wave evolves over time, not taking special relativity into account.

Special theory of relativity The special theory of relativity is based on the principle that the speed of light is the same for every uniformly moving observer. Together with general relativity and quantum theory, it forms the foundation of modern physics.

Speed of light The speed of light c is almost 300,000 km/s in vacuum. This constant of nature plays a central role in the theory of special relativity, where it links space and time as well as masses, momenta, and energies.

Spin Quantum mechanical analogue of the angular momentum of a classical object. Bosons have integer spin (including 0), fermions have half-integer spin (e.g., 1/2 or 3/2).

Standard Model of particle physics The Standard Model provides the fundamental basis for all physics known today with the exception of gravity. It explains the known matter (and antimatter) and its fundamental interactions (electromagnetic, weak, and strong interactions) using six quarks, six leptons (electron, muon, tauon, and related neutrinos), and the interaction particles (photon, gluon, W and Z bosons). The description is based on the rules of relativistic quantum field theory. Quantum electrodynamics (QED), which Feynman co-founded, is part of the Standard Model, as is quantum chromodynamics (QCD). Quantum gravity, on the other hand, is not part of the Standard Model.

Strangeness The strangeness of a particle indicates whether it contains strange quarks or antiquarks, and is defined as the difference between the number of strange antiquarks and strange quarks. Originally, strangeness was introduced before the discovery of quarks to explain, e.g., the relatively long lifetime of kaons: only the weak interaction is able to change the strangeness in a decay.

String theory String theory is based on the idea that the elementary objects of nature are not point-like particles, but tiny strings. Interestingly, string theory automatically contains gravitons and thus a quantum theory of gravity.

Strong interaction The strong interaction describes the influences that quarks and gluons exert on each other due to their color charge. Since protons and neutrons consist of quarks, the strong interaction also leads to strong attractive forces between these particles, which bind them closely together in the atomic nuclei. It is therefore also referred to as the strong nuclear force.

Superfluidity When liquid helium behaves at very low temperatures like a liquid without any viscosity, this is called superfluidity. Such a superfluid liquid can penetrate even the smallest pores and flow without any flow resistance. Feynman was able to explain this behavior with the help of Bose–Einstein condensation, i.e., with the tendency certain atoms have to occupy the same macroscopic quantum ground state.

Tauon From today's point of view the tauon is an elementary point-like particle with negative charge and spin 1/2. It thus has similar properties to the electron and the muon, but is about 3500 times heavier than the electron and is therefore unstable.

Time dilation This is the effect where an observer viewing a moving clock will see it running more slowly from his point of view than if it were stationary. The Lorentz factor indicates the extent of this time dilatation, which is a consequence of the special theory of relativity.

Uncertainty principle The Heisenberg uncertainty principle states that the location and the momentum of a particle cannot both be known with arbitrary accuracy at the same time. The product of the uncertainties in these two quantities cannot be pushed below a certain value given by the Planck constant. The more precisely one knows the location of a particle, the less precisely its momentum is known, and vice versa. For energy and time, there is an analog uncertainty relation which allows the existence of virtual particles.

Vertex In a Feynman diagram, a vertex is a point at which one or more new particles (e.g., a photon) are created or destroyed.

W bosons Positively or negatively charged particles which, together with the neutral Z boson, mediate the weak interaction. Unlike the massless photon which mediates the electromagnetic interaction, the W^+ and the W^- bosons are very heavy (about eighty times heavier than the proton), which explains the extremely short range of the weak interaction. W bosons can convert different particles into each other and thus cause, e.g., radioactive beta decay.

Wave function Another name for a quantum wave.

Wavelength If we look at a wave at a fixed time, the wavelength is the distance between two adjacent wave crests or troughs. In quantum mechanics, the wavelength of a quantum wave determines the momentum of the associated particle.

Weak interaction The weak interaction describes the influence of the W and Z bosons on quarks and leptons within the framework of the Standard Model. It can convert different types of quarks and leptons into each other and thus triggers many particle decays, such as radioactive beta decay.

Z boson Very heavy neutral particle which, together with the charged W bosons, mediates the weak interaction.

Sources and Literature

Richard Feynman was a kind of pop star in physics in the middle and late twentieth century. There is a wide range of books, texts, and internet videos by him and about him. The following sources constitute only an incomplete selection, but they are ones that have been a great help in writing this book.

Biographies and Anecdotes

Laurie M. Brown (Editor), John S. Rigden (Editor): *Most of the Good Stuff: Memories of Richard Feynman*, American Institute of Physics 1993

Richard P. Feynman: *Los Alamos From Below: Reminiscences 1943–1945*, on the Internet, e.g., at http://calteches.library.caltech.edu/34/3/FeynmanLosAlamos.htm and at youtube.com

Richard P. Feynman: *Surely You're Joking, Mr Feynman!* Norton & Co. 2018 (it can also be found on the Internet)

Richard P. Feynman: *The Pleasure of Finding Things Out*, Basic Books 2005 (it can also be found on the Internet; the video *BBC Horizon – Richard P. Feynman – The Pleasure of Finding Things Out* is available at https://www.dailymotion.com/video/x24gwgc)

Richard P. Feynman: *What Do You Care What Other People Think?* Norton & Co. 2001 (it can also be found on the Internet)

Feynman's letter of 17 October 1946 to his deceased wife Arline can be found at www.lettersofnote.com (search for *Arline* or *I love my wife. My wife is dead.*)

Murray Gell-Mann: *Dick Feynman – The Guy in the Office down the Hall*, Physics Today (February 1989), http://www.physics.rutgers.edu/~jwolf/Gell-mann_Feynman.pdf

James Gleick: *Genius: The Life and Science of Richard Feynman*, Pantheon 1992

John and Mary Gribbin: *Richard Feynman: A Life in Science*, Dutton Adult 1997

J. Resag, *Feynman and His Physics*, Springer Biographies,
https://doi.org/10.1007/978-3-319-96836-0

Charles Hirshberg: *My Mother, the Scientist*, Popular Science May 2002, on the Internet for example at http://www.aas.org/cswa/status/2003/JANUARY2003/MyMotherTheScientist.html
George Johnson: *The Jaguar and the Fox*, an article about Gell-Mann and Feynman from July 2000, at http://www.theatlantic.com
Lawrence M. Krauss: *Quantum Man: Richard Feynman's Life in Science*, Norton & Co. 2011
Ralph Leighton: *Tuva or Bust! Richard Feynman's Last Journey*, Norton & Co. 2000
Jagdish Mehra: *Climbing The Mountain: The Scientific Biography of Julian Schwinger*, Oxford University Press 2003
Jagdish Mehra: *Meine letzte Begegnung mit Richard Feynman*, Physikalische Blätter, Band 44, 1988, S. 146–148, https://onlinelibrary.wiley.com/doi/10.1002/phbl.19880440509/abstract
Jagdish Mehra: *The Beat of a Different Drum: Life and Science of Richard P. Feynman*, Clarendon Press 1994
Silvan S. Schweber: *Feynman and the visualization of space-time processes*, http://www.rpgroup.caltech.edu/courses/aph105c/2006/articles/Schweber1986.pdf
Silvan S. Schweber: *QED and the Men Who Made It: Dyson, Feynman, Schwinger and Tomonaga*, Princeton University Press 1994
Dennis Silverman: *Richard Feynman, A Life of Many Paths*, http://www.physics.uci.edu/~silverma/RichardFeynman.pdf

Lectures and Talks

Feynman, Leighton, Sands, *The Feynman Lectures on Physics*, Basic Books 2011, also on the internet at http://www.feynmanlectures.caltech.edu/
Feynman's 1965 Nobel Lecture *The Development of the Space-Time View of Quantum Electrodynamics* (as well as Schwinger's and Tomonaga's lectures) can be found at http://www.nobelprize.org/nobel_prizes/physics/laureates/1965/
Richard P. Feynman: *QED: The Strange Theory of Light and Matter*, Princeton University Press 2014 (based on the video *The Sir Douglas Robb Lectures, University of Auckland*, 1979, http://www.vega.org.uk/video/subseries/8, also to be found at *Richard Feynman Lecture on Quantum Electrodynamics: QED* at youtube.com; particularly nice is *You don't like it? Go somewhere else! by Richard Feynman, the QED Lecture at University of Auckland* at https://www.youtube.com/watch?v=iMDTcMD6pOw)
Richard P. Feynman: *The Character of Physical Law*, The MIT Press 2017 (also on the Internet at http://people.virginia.edu/~ecd3m/1110/Fall2014/The_Character_of_Physical_Law.pdf; the original videos of the *Richard Feynman Messenger Lectures* from 1964 are available at http://research.microsoft.com/apps/tools/tuva/ and http://www.cornell.edu/video/playlist/richard-feynman-messenger-lectures and at youtube.com).
Richard P. Feynman: *What is Science?* The Physics Teacher 1969, on the Internet at http://www.fotuva.org/feynman/what_is_science.html

Richard P. Feynman, Steven Weinberg: *Elementary Particles and the Laws of Physics: The 1986 Dirac Memorial Lectures*, Cambridge University Press 1999; Feynman's presentation *The reason for antiparticles* is also available as a video on youtube.com and partly as a transcript on the Internet, e.g., at http://assets.cambridge.org/97805216/58621/excerpt/9780521658621_excerpt.pdf
Richard P. Feynman: *There's Plenty of Room at the Bottom*, also on the Internet, e.g., at http://www.its.caltech.edu/~feynman/plenty.html
Richard Feynman: *Infinitesimal Machinery*, Journal of Microelectromechanical Systems, Vol. 2, No. I, March 1993, http://maecourses.ucsd.edu/~pbandaru/mae268-sp09/Class%20Readings_files/feynman_infinetismal%20machinery.pdf
Richard P. Feynman: *Quantum Theory of Gravitation*, Acta Physica Polonica 24 (1963): 697–722, e.g., at http://blogs.umass.edu/grqft/files/2014/11/Feynman-gravitation.pdf
Richard P. Feynman: *Feynman Lectures On Computation*, Addison-Wesley 1996
W. Daniel Hillis: *Richard Feynman and The Connection Machine*, Physics Today 42(2), 78 (1989), also on the Internet
Henry W. Kendall: *Deep Inelastic Scattering: Experiments on the Proton and the Observation of Scaling*, Nobel Lecture (1990), https://www.nobelprize.org/nobel_prizes/physics/laureates/1990/kendall-lecture.html
Fernando B. Moringo, David Pines, Richard P. Feynman: *Feynman Lectures on Gravitation*, Westview Press, also at http://www.download-genius.com/download-k:Feynman%20Lectures%20on%20Gravitation.html?aff.id=6253, preface by John Preskill and Kip S. Thorne at http://www.theory.caltech.edu/~preskill/pubs/preskill-1995-feynman.pdf
Stephen Wolfram: *A Short Talk about Richard Feynman* (2005), http://www.stephenwolfram.com/publications/short-talk-about-richard-feynman/

Interviews

Interviews with Freeman Dyson, Hans Bethe, John Wheeler, and Murray Gell-Mann can be found at www.webofstories.com
Joan Feynman: *The Aurora* and other interviews with her on youtube.com
Richard Feynman: *Fixing Radios*, on youtube.com
Murray Gell-Mann talks about Richard Feynman, on youtube.com
Interview with George Zweig by Panos Charitos, EP department (CERN 2013), http://ep-news.web.cern.ch/content/interview-george-zweig

Scientific Articles and Books

John C. Baez and Emory F. Bunny: *The Meaning of Einstein's Equation*, http://math.ucr.edu/home/baez/einstein/einstein.pdf
Sébastien Balibar: *Looking Back at Superfluid Helium*, Séminaire Poincaré 1 (2003) 11–20, e.g., at http://www.bourbaphy.fr/balibar.pdf

H.A. Bethe: *The Electromagnetic Shift of Energy Levels*, Phys. Rev. 72, 339 (1947), http://web.ihep.su/dbserv/compas/src/bethe47/eng.PDF

Laurie M. Brown (Editor): *Selected Papers of Richard Feynman with Commentary*, World Scientific Series in 20th Century Physics, Vol. 27 (2000)

Freeman J. Dyson: *The Radiation Theories of Tomonaga, Schwinger, and Feynman*, Phys. Rev. 75, 486 (February 1949), e.g., at http://web.ihep.su/dbserv/compas/src/dyson49b/eng.pdf

Freeman J. Dyson: *The S Matrix in Quantum Electrodynamics*, Phys. Rev. 75, 1736 (June 1949)

A. Einstein: *Quantentheorie des einatomigen idealen Gases – Zweite Abhandlung*. In: Sitzungsberichte der preussischen Akademie der Wissenschaften. 1925, S. 3–10, e.g., at https://web.physik.rwth-aachen.de/~meden/boseeinstein/einstein1925.pdf, English translation *Quantum Theory of a Monoatomic Ideal Gas*, e.g., at http://www.thphys.uni-heidelberg.de/~amendola/otherstuff/einstein-paper-v2.pdf

Richard P. Feynman: *Forces and stresses in molecules*, Bachelor thesis at the Massachusetts Institute of Technology 1939, http://hdl.handle.net/1721.1/10786, also published as *Forces in molecules*, Physical Review, Vol. 56, 1939, S. 340–343

Richard P. Feynman: *Principles of least action in quantum mechanics*, Doctoral thesis at Princeton University 1942, http://cds.cern.ch/record/101498/files/Thesis-1942-Feynman.pdf

Richard P. Feynman: *Space-Time Approach to Non-Relativistic Quantum Mechanics*, Rev. Mod. Phys. 20, 367 (April 1948), e.g., at http://www.fafnir.phyast.pitt.edu/py3765/PathIntegral.pdf

Richard P. Feynman: *The Theory of Positrons*, Phys. Rev. 76, 749 (September 1949), e.g., at http://authors.library.caltech.edu/3520/1/FEYpr49b.pdf

Richard P. Feynman: *Space-Time Approach to Quantum Electrodynamics*, Phys. Rev. 76, 769 (September 1949), e.g. at http://authors.library.caltech.edu/3523/1/FEYpr49c.pdf

Richard P. Feynman: *Mathematical Formulation of the Quantum Theory of Electromagnetic Interaction*, Phys. Rev. 80, 440 (November 1950)

Richard P. Feynman: *An Operator Calculus Having Applications in Quantum Electrodynamics*, Phys. Rev. 84, 108 (October 1951)

Richard P. Feynman: *Atomic Theory of the λ Transition in Helium*, Phys. Rev. 91, 1291 (September 1953), e.g. at http://authors.library.caltech.edu/3537/1/FEYpr53b.pdf

Richard P. Feynman: *Atomic Theory of Liquid Helium Near Absolute Zero*, Phys. Rev. 91, 1301 (September 1953)

Richard P. Feynman: *Atomic Theory of the Two-Fluid Model of Liquid Helium*, Phys. Rev. 94, 262 (April 1954), e.g. at https://core.ac.uk/download/files/200/4872090.pdf

Richard P. Feynman: *Application of Quantum Mechanics to Liquid Helium*, Progress in Low Temperature Physics 12/1955

Richard P. Feynman, Michael Cohen: *Energy Spectrum of the Excitations in Liquid Helium*, Phys. Rev. 102, 1189 (June 1956)

Richard P. Feynman, M. Gell-Mann: *Theory of the Fermi Interaction*, Phys. Rev. 109, 193 (January 1958), e.g., at http://authors.library.caltech.edu/3514/1/FEYpr58.pdf

Richard P. Feynman: *Simulating Physics with Computers*, International Journal of Theoretical Physics Vol. 21 (1982), PDF also at https://people.eecs.berkeley.edu/~christos/classics/Feynman.pdf

Francis Halzen, Alan D. Martin: *Quarks and Leptons: An Introductory Course in Modern Particle Physics*, Wiley 1984, PDF also at http://ajbell.web.cern.ch/ajbell/Documents/eBooks/Quarks%20&%20Leptons.pdf

Tony Hey: *Richard Feynman and Computation*, Contemporary Physics Vol. 40 (1999), http://www.cs.indiana.edu/~dgerman/hey.pdf

Tsung-Dao Lee, C. N. Yang: *Question of Parity Conservation in Weak Interactions*, Phys. Rev. 104, 254 (October 1956), e.g., at http://www2.pv.infn.it/~fontana/download/lect/LeeYang.pdf

Tsung-Dao Lee, C. N. Yang: *Parity Nonconservation and a Two-Component Theory of the Neutrino*, Phys. Rev. 105, 1671 (March 1957)

Alfred Leitner's Old Physics Stories, http://alfredleitner.com/, especially the video *Superfluid Liquid Helium (Isotope 4)* (39 min., 1963)

Alexander Lesov: *The Weak Force: From Fermi to Feynman*, http://arxiv.org/pdf/0911.0058.pdf

Chris Toumey: *Reading Feynman into Nanotechnology: A Text for a New Science*, https://web.archive.org/web/20090919171108/http://scholar.lib.vt.edu/ejournals/SPT/v12n3/toumey.pdf

William F. Vinen: *The Physics of Superfluid Helium*, http://cds.cern.ch/record/808382/files/p363.pdf

John A. Wheeler, Richard P. Feynman: *Interaction with the Absorber as the Mechanism of Radiation*, Rev. Mod. Phys. 17, 157 (April 1945), http://authors.library.caltech.edu/11095/

John A. Wheeler, Richard P. Feynman: *Classical Electrodynamics in Terms of Direct Interparticle Action*, Rev. Mod. Phys. 21, 425 (July 1949), http://authors.library.caltech.edu/48015/

Chien-Shiung Wu, E. Ambler, R. W. Hayward, D. D. Hoppes, R. P. Hudson: *Experimental Test of Parity Conservation in Beta Decay*, Phys. Rev. 105, 1413 (February 1957), e.g., at https://journals.aps.org/pr/pdf/10.1103/PhysRev.105.1413

H.-Dieter Zeh: *Feynman's Interpretation of Quantum Theory*, Eur. Phys. J. H https://doi.org/10.1140/epjh/e2011-10035-2, arxiv:0804.3348v6, https://arxiv.org/ftp/arxiv/papers/0804/0804.3348.pdf